PIONEERS TO THE PAST

James Henry Breasted (center) with Sheikh Suwan of es-Sabkha and his Grandon, May 1920

PIONEERS TO THE PAST

AMERICAN ARCHAEOLOGISTS IN THE MIDDLE EAST
1919–1920

edited by

GEOFF EMBERLING

THE ORIENTAL INSTITUTE MUSEUM PUBLICATIONS • NUMBER 30
THE ORIENTAL INSTITUTE OF THE UNIVERSITY OF CHICAGO

Library of Congress Control Number: 2009941795
ISBN-10: 1-885923-70-8
ISBN-13: 978-1-885923-70-7

The Oriental Institute, Chicago

This volume has been published in conjunction with the exhibition
Pioneers to the Past: American Archaeologists in the Middle East, 1919–1920,
presented at The Oriental Institute Museum, January 12 to August 29, 2010.

Oriental Institute Museum Publications No. 30

The Series Editors Leslie Schramer and Thomas G. Urban would like to thank
Professor McGuire Gibson, Thomas R. James, Adrienne Frie, Paula Manzuk,
and Travis Garth for their help in the production of this volume.

Published by The Oriental Institute of the University of Chicago
1155 East 58th Street
Chicago, Illinois 60637 USA
oi.uchicago.edu

Duo-tones prepared by Nimrod Systems, Ltd., Chicago, Illinois

Printed by Chicago Press Corporation, Chicago, Illinois

The paper used in this publication meets the minimum requirements of American National Standard for
Information Service — Permanence of Paper for Printed Library Materials, ANSI Z39.48-1984
∞

TABLE OF CONTENTS

CONTRIBUTORS

ABOUT THE CONTRIBUTORS:

- **Orit Bashkin** is Assistant Professor of Middle Eastern History in the Department of Near Eastern Languages and Civilizations at the University Chicago, specializing in the history of Iraq. She is the author of *The Other Iraq: Pluralism and Culture in Hashemite Iraq.*
- **Geoff Emberling** is Museum Director at the Oriental Institute and curator of the Pioneers to the Past exhibit. He is the co-editor of *Catastrophe: The Looting and Destruction of Iraq's Past.*
- **James L. Gelvin** is Professor of History at the University of California, Los Angeles. Among his books are *The Modern Middle East: A History* and *Divided Loyalties: Nationalism and Mass Politics in Syria at the Close of Empire.*
- **Morag M. Kersel** is a Postdoctoral Fellow at the Joukowsky Institute for Archaeology and the Ancient World at Brown University.
- **John A. Larson** is the Archivist at the Oriental Institute and an acknowledged expert on the life and work of James Henry Breasted.
- **Emily Teeter** is a Research Associate (Egyptology) and Coordinator of Special Exhibits at the Oriental Institute of the University of Chicago.

IMAGE CREDITS:

Figures unless otherwise credited are from the Oriental Institute Archives: 4.3–4, 4.6, 4.9–13, 4.15, 4.18, 4.60 — Anna Ressman; 4.16–17 — Art Institute of Chicago; map by Leslie Schramer

FOREWORD

GIL J. STEIN
DIRECTOR, ORIENTAL INSTITUTE

The Oriental Institute's special exhibit Pioneers to the Past: American Archaeologists in the Middle East, 1919–1920, highlights the interconnected stories of an important figure in intellectual history, James Henry Breasted, the beginnings of American scientific archaeology in the Near East at a crucial turning point in world history, and the birth of the modern Middle East.

James Henry Breasted was one of the most remarkable individuals in the history of American scholarship on the ancient Near East. He was a brilliant teacher, Egyptologist, communicator, and above all an institution-builder who transformed the fundamental character of research on the origins of civilization. Breasted's greatest achievement was the founding of the Oriental Institute at the University of Chicago in 1919, through the generous support of John D. Rockefeller Jr. The Oriental Institute embodies Breasted's vision of an inter-disciplinary research center that unites archaeology, textual studies, and art history as three complementary methodologies to provide a holistic understanding of ancient Near Eastern civilizations.

The founding of the Oriental Institute can only be understood in the context of the radically transformed political landscape of the Near East at the end of World War I. By 1919, the Ottoman Empire had collapsed, and the victorious English and French allies had been awarded "mandates" of non-colonial administrative control over most of the Middle East (with the exception of Turkey and Iran). Thus, for the first time, almost all of what Breasted called the "Fertile Crescent" was now accessible to Western archaeologists and philologists seeking to explore the ruins and textual records that would form the basis for understanding life in the cradle of ancient Near Eastern civilization. It was also an unparalleled opportunity for the University of Chicago to play a leading research role in a discipline and a region where research had traditionally been dominated by European scholars and institutions.

Breasted immediately grasped the significance of this historical moment and approached John D. Rockefeller Jr. with an ambitious and detailed proposal for financial support to found the Oriental Institute as an interdisciplinary center for archaeological and philological research on the ancient Near East and its role in the origins of Western civilization. In May 1919, Rockefeller wrote back to Breasted agreeing to fund the Oriental Institute, and just three months later — in August 1919 — Breasted and a small team of scholars set sail for the Near East on what would be an eleven-month odyssey across the region in 1919–1920. Breasted's explicit goal was to establish good working relationships with the mandatory governments of the region, while identifying the most significant ancient urban centers across the region, so that the Oriental Institute could gain permission to conduct large-scale, long-term projects of archaeological research at these sites.

The fascinating mix of politics, scholarship, and history (both ancient and modern), as seen through a focus on the larger-than-life persona of James Henry Breasted lies at the heart of our special exhibit Pioneers to the Past. Breasted's letters and photographs from his trip provide a window into the engagement of modern scholarship with the ancient world, in a highly charged setting of power politics in the early twentieth century. The essays in this catalogue explain the historical, legal, and political context in a way that greatly enriches our understanding of Breasted's journey and its aftermath. Geoff Emberling and Emily Teeter have done a wonderful job in bringing this little-known, but crucial, period of history to life.

FIGURE 1.1 James Henry Breasted, ca. 1933

1. INTRODUCTION

GEOFF EMBERLING
MUSEUM DIRECTOR, ORIENTAL INSTITUTE

James Henry Breasted (figs. 1.1–2) was a brilliant, determined, and energetic scholar whose vision for the field of ancient Middle Eastern[1] (or "Oriental") studies transformed scholarship in the early twentieth century and continues to shape the way archaeologists and historians understand the civilizations of Egypt, Mesopotamia, and their neighbors. The first American to receive a PhD in Egyptology, Breasted's scholarship encompassed monumental surveys of Egyptian history, contributions to understanding of Egyptian language, and popular books that reached an astonishingly wide international audience. He argued that the origins of Western civilization were to be sought further back in time (and farther east) than the classical world, and for the relevance of ancient Middle Eastern civilizations to Europe and America. He also argued for the interconnectedness of these cultures along the Fertile Crescent — a term he coined to describe the arc of fertile agricultural land extending from Mesopotamia, across the northern Mesopotamian plains, and down the Mediterranean coast — and for the necessity of interdisciplinary study to understand them. But perhaps Breasted's greatest legacy to the field was founding the Oriental Institute, a research institute at the University of Chicago that today remains at the center of study of the archaeology, art, languages, and history of the ancient Middle East.

This volume, and the exhibit for which it was written, follows Breasted's inaugural expedition for the Institute in 1919–1920 as he and four companions traveled through Egypt and what are now Iraq, Syria, Lebanon, and Israel. The aims of the trip were to purchase antiquities for study and display in Chicago and to prepare for excavations by meeting with colleagues and officials across the Middle East and by identifying sites. The volume's title — *Pioneers to the Past: American Archaeologists*

---

[1] In Breasted's time, the region was called the Near East, and scholars today still refer to the area of study as the ancient Near East. In light of common popular usage today, we generally use "ancient Middle East" in this catalogue, but the geographical referent is the same.

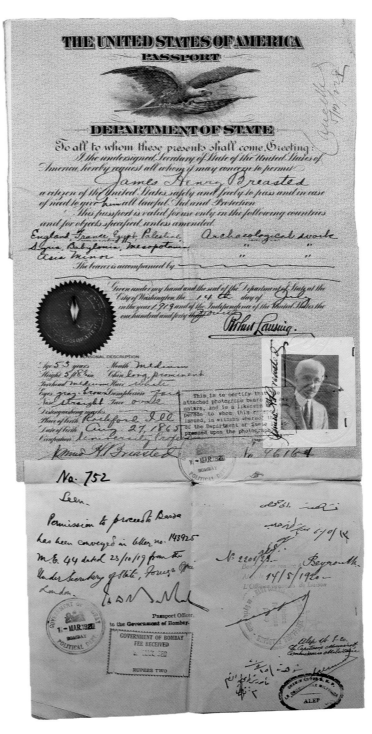

FIGURE 1.2 The first page of the passport James Henry Breasted used on his 1919–1920 journey

in the Middle East, 1919–1920 — evokes several different aspects of the trip. Breasted and his group were among the first American archaeologists to travel to and work in the Middle East, and Breasted was very conscious of his role in putting America (and particularly Chicago) on the world stage as holders of significant collections of art and artifacts from the Middle East. And because Breasted and his team were fundamentally more interested in the past than in the present reality of the region, there is a very real sense in which their trip was to a past landscape. Finally, images of the expedition's wagons look like nothing more than pioneers conquering the great American West, except that they are situated along the Euphrates River in what is now Syria.

Much of the trip itself was a real-life adventure story, which is to say that parts were thrilling, while others were dull, uncomfortable, frustrating, or scary. The Middle East in 1919–1920 was extremely unsettled in the immediate aftermath of World War I; military operations, nationalist movements, lawlessness, and tribal revolts all endangered the expedition at one point or another. Once the team left the areas under British military control, food could be scarce and unappealing and sleep could be hard to come by. Yet for Breasted, the trip allowed him to visit the great monuments of Assyria, Babylonia, Persia, Phoenicia, and Israel that he had read about (indeed, written about) but had never seen.

For Breasted personally, the trip was alternately exciting, difficult, and deeply lonely. He turned 54 just as the trip began, and had left his wife Frances, his 21-year-old son Charles, 11-year-old son James Jr., and 5-year-old daughter Astrid at home for what would turn out to be an eleven-month journey covering more than 20,000 miles. Many of his letters begin with accounts of receiving (or not receiving) letters from home. As he wrote to Frances:

> I hope you may never be in a situation where letters mean so much and are looked for with such eagerness and longing. All this is part of the price to be paid for results, and my compensation lies in the conviction, right or wrong, that it is service to science — nothing great or brilliant — but the best I am able to offer, and done as I feel, at a cost to be measured only by the extreme sensitiveness, loneliness and almost morbid love of home with which I am unfortunately encumbered. – *JHB to Frances, April 12, 1920*

The extended separation was naturally difficult for Frances as well. As their son Charles later wrote, she followed the journey through Breasted's letters home, putting pins in a map to trace his route, but her anxieties and resentments grew to such an extent that she became physically ill just as Breasted arrived home, and was unable to meet him at the train station (C. Breasted, *Pioneer to the Past*, p. 315).

As one might expect during a tiring journey, Breasted occasionally expressed exasperation with his traveling companions:

> I sometimes or perhaps I should say <u>most</u> times seem to be in charge of a <u>kindergarten party</u>, absolutely dependent on me for nearly everything they need except the breath they breathe. It is better for me not to put down here all that is said and done, but there has been no scene and no row, for I simply will not have that kind of thing in my expedition. It is far better afterward to have been inexhaustibly patient. – *JHB to Frances, April 18, 1920*

If anything, the surprise is that he did not express such feelings more often, given the difficulty of the conditions they encountered along the way.

The surviving sources for this journey include more than fifty long letters that Breasted wrote back to his family, to donors, to President Judson of the University of Chicago, and others, as well as more than 1,100 photographs. These sources are all preserved in the Archives of the Oriental Institute.

Several accounts of the expedition have been published, including Breasted's own summaries; a book by one of his companions, William Shelton; a version comprised of extensive quotes from Breasted's letters in the biography of Breasted written by his son, Charles; and more recently, a selection of photographs published in microfilm with a summary by Oriental Institute docent Ruth Marcanti. The trip is also discussed in a forthcoming biography of Breasted written by Jeffrey Abt.

FIGURE 1.3 The tympanum over the main entrance of the Oriental Institute, entitled "East Teaching the West." The allegorical composition shows the East, depicted as an Egyptian scribe, gesturing to the West, represented by a young man draped in a cloak. In his hands the West holds a plaque inscribed in hieroglyphs, which reads "We behold your goodness." People and things emblematic of ancient (Middle Eastern) and modern (Western) civilizations flank the central figures on the left and right

The Oriental Institute exhibit and this volume differ from previous accounts by including a selection of unpublished photographs, letter excerpts, and other documents related to the trip from the Archives of the Oriental Institute. It also provides a broader context for the Breasted expedition by exploring some of the many issues raised by the journey. These issues include ownership of the past, the antiquities trade, links between past civilizations and modern nations, the political importance of archaeology, and cultural relationships between the United States and the Middle East. Each of these issues continues to be discussed by archaeologists and museum professionals to this day.

Breasted's argument that Western civilization began in the Middle East is historically accurate and today is not in dispute. It is clear, however, that Breasted thought that civilization had passed from the Middle East to the West, a view enshrined in the tympanum over the main entrance of the Oriental Institute (fig. 1.3). He also thought that the inhabitants of the region of that time were not worthy heirs of the great ancient civilizations. As he put it:

> the number of educated Egyptians who can appreciate such things [ancient Egyptian objects] is an insignificant handful, while on the other hand, as our birthright and inheritance from the past, Egypt can be a wonderful educational influence in civilized lands of the West. – *JHB to Charles, September 25, 1919*

Breasted thus employed his historical argument as a justification for removing objects from Egypt and bringing them to museums in the West.

Since Breasted's time, the antiquities trade has become the focus of serious debates among archaeologists and museums throughout the world, as many have seen direct connections between the purchase of antiquities and the destruction of archaeological sites. After the 1970 UNESCO Convention on the illegal transfer of cultural property, some museums stopped purchasing antiquities that were not demonstrably found before 1970, and more recently (and reluctantly), larger museums have begun to follow suit. The practice of giving foreign archaeological teams a share of their finds (also called "division" or *partage*), which was common in Breasted's day, has long since ended as a regular practice — field projects today normally leave their finds in the host country.

Middle Eastern countries, particularly Egypt and Turkey, have made efforts to force the return of some objects, like the Lydian silver hoard from Turkey and the famous bust of Nefertiti from Egypt. At the same time, some American and European museum directors have worked to develop new justifications for display (and ownership) of ancient Middle Eastern art. For example, James Cuno, Director of the Art Institute of Chicago, makes the highly controversial argument that modern nation-states and their inhabitants do not have an inherent connection to the ancient cultures that once lived within their borders, and that the interests of the international community are best served by the movement of art across those borders. These large questions about the ownership of the past and its connection to present-day political and cultural realities thus continue to be debated today.

One of the striking things about Breasted's journey is the access that he was given to the highest levels of military and political authority wherever he traveled. This access was granted because archaeology and history had a political importance for colonial officials. Control of an area required intellectual control of its past and of the political uses that could be made of it. Thus the rising nationalism across the Middle East in the 1920s also seized on antiquities as a potent symbol of national identity.

Archaeologists working in the Middle East today rarely have contacts at the highest level of their host government — although it does happen — but political struggles over the meaning of the past continue. As just one example, one could mention the ways in which archaeology has been used in different ways to argue for territorial rights of Israelis and Palestinians.

One aspect of the sources that is difficult to avoid is the occasional but overt expressions of racism expressed by Breasted. In this, he was certainly a man of his time — Europeans (and increasingly Americans) had developed ideas of superiority as a result of the great social and economic advancements of the previous half century. It is a poignant comment on the power of the colonial system that Breasted could on the one hand argue for the importance of the Middle East in human history at the same time that he denied the connection of the great ancient civilizations to the modern inhabitants of the region.

American and European archaeologists have certainly changed in our personal attitudes toward the Middle East. Yet in some ways, the normal research model for archaeological projects remains unchanged from Breasted's time. American or European archaeologists define a research agenda, carry out field research, and (all too often) take the information home without sharing results in significant ways with host countries. But in fact there are many positive steps toward more collaborative work with our Middle Eastern colleagues, from having co-directors on excavations, outreach to local communities, training of archaeologists and museum staff, training of local workers on useful computer and language skills, to help with conservation of monuments and construction and installation of local museums. Changes in the normal model of research also continue to be a point of discussion among archaeologists.

The following chapters provide a variety of perspectives on Breasted's journey. *Archaeology in the Middle East before 1920: Political Contexts, Historical Results* provides an overview of what was known about ancient Middle Eastern societies at the time of the expedition. It also reviews the varying financial and institutional support for archaeology as a way of understanding the political importance of understanding the past. *The Middle East Breasted Encountered, 1919–1920*, by James L. Gelvin, reviews the history of the Middle East during and immediately after World War I with particular reference to the chronology and itinerary of Breasted's journey, as well as to the impact of larger political debates on archaeological practice. *The First Expedition of the Oriental Institute, 1919–1920*, by Geoff Emberling and Emily Teeter, presents the voyage itself in detail with quotes from Breasted's letters and selections from the photographic archives. *The Changing Legal Landscape for Middle Eastern Archaeology in the Colonial Era, 1800–1930*, by Morag M. Kersel, discusses the connections between colonial administrations, Middle Eastern nationalism, and developing antiquities laws that sought to make a claim to the ancient past of the region. *The Arab Revival, Archaeology, and Ancient Middle Eastern History*, by Orit Bashkin, is a unique contribution to the history of archaeology in the Middle East, discussing some ways in which the educated elite in the Middle East viewed the activities of foreign archaeologists. An epilogue by Emily Teeter discusses the aftermath of the expedition, from Breasted's recommendations to the University of Chicago about the future activities of the Institute, and some of the ways in which the Institute in the 1920s and afterward implemented or deviated from Breasted's plan. Appendices make available some key documents. Throughout, Breasted's spellings of names are maintained in direct quotes, but an attempt has been made to use modern versions elsewhere in the text.

A number of activities are planned in conjunction with the Oriental Institute exhibit (January 12–August 29, 2010). A companion display at the Art Institute of Chicago highlights objects Breasted bought for that museum. A searchable final transcript of Breasted's letters will be published in an online-only form in a new Oriental Institute Archives series. The Breasted biography written by his son Charles (*Pioneer to the Past: The Story of James Henry Breasted, Archaeologist*) will be reprinted with a selection of photographs that were not included in the original publication. Finally, all the photos taken on the trip will be made available in searchable form online.

ACKNOWLEDGMENTS

It is a pleasure to thank the many people who have made the exhibit and catalogue possible.

The Special Exhibits program at the Oriental Institute is generously supported by Exelon Corporation. Pioneers to the Past has also been supported by private donors including Barbara Breasted Whitesides and Peggy Grant.

Thanks to Gil Stein, Director of the Oriental Institute, for his consistent support of the Special Exhibits program and for his idea that Breasted's expedition was due for renewed attention.

For help in developing the content of the exhibit, I would like first to thank John Larson, Oriental Institute Archivist, whose work in arranging the scanning of photographs, collating image captions, managing the paper records from the expedition, and arranging for transcription of Breasted's letters were the essential foundation of the exhibit. His deep knowledge of the material has been a great resource throughout the process. John has been assisted by a number of dedicated volunteers, including Peggy Grant, who transcribed Breasted's letters from the field.

Emily Teeter, Oriental Institute Research Associate and Coordinator of Special Exhibits, has been involved at every stage of exhibit preparation, from working with the museum's Community Focus Group to testing aspects of exhibit presentation, to selection of purchased objects, and writing significant portions of the catalogue and the audio guide script.

Thanks to James Gelvin, Morag Kersel, and Orit Bashkin for their excellent and timely contributions to the catalogue and for their comments on other chapters. Thanks too to Adam Smith and to an anonymous reviewer for comments; as usual, none of the commentators can be held responsible for any shortcomings in the catalogue.

Tom Urban and Leslie Schramer in the Oriental Institute Publications office shepherded the catalogue through publication with professionalism and good humor.

We have been fortunate to continue to have exhibit design assistance from Dianne Hanau-Strain of Hanau-Strain Associates.

Within the Oriental Institute Museum, Carole Krucoff, Head of Museum Education and Public Programs, represented the interests of the museum-going public with her usual intelligence, determination, and tact. Wendy Ennes took the lead in linking the exhibit and catalogue to high-school history curricula. Head Preparator Erik Lindahl and Preparator Brian Zimerle led the design of the exhibit. Tom James, Assistant Curator of Digital Collections, designed an interface for the interactive presentation of photos in the gallery, and managed the photographs through the curation of the exhibit and production of this catalogue. Alison Whyte provided conservation assessments of the objects, Anna Ressman took new photographs on the usual compressed schedule, and Helen McDonald and Susan Allison kept track of the objects as they moved around the museum. Finally, Adrienne Frie provided a great boost of careful editing and enthusiasm for all aspects of the exhibit.

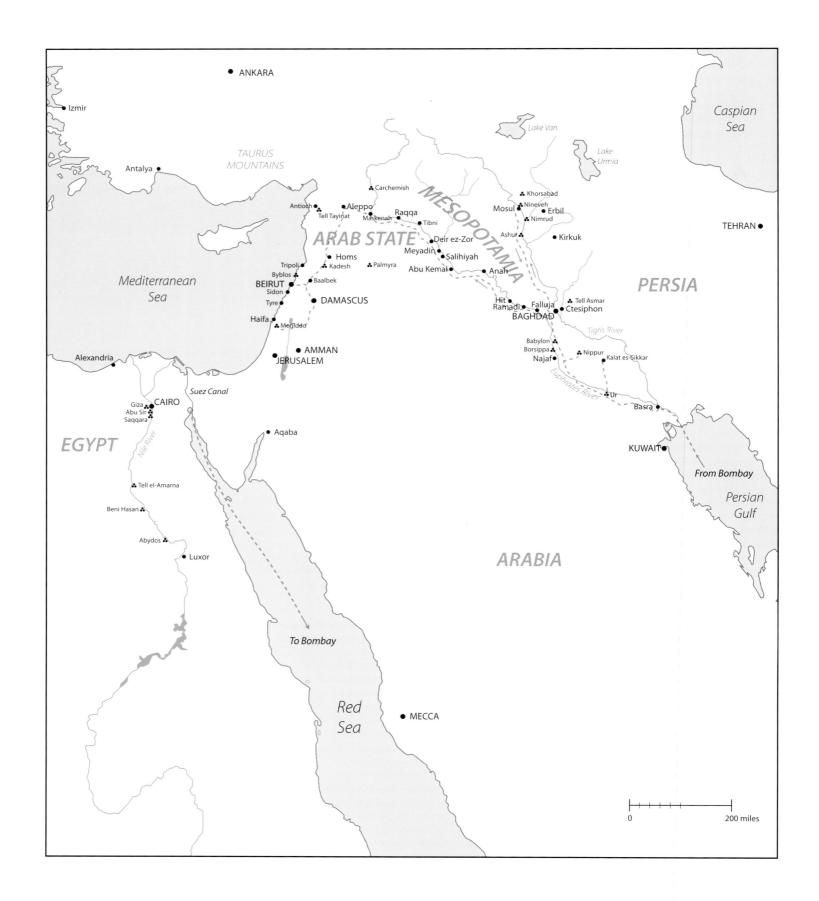

Map showing the route (orange line) of the Oriental Institute Expedition of 1919–1920

2. ARCHAEOLOGY IN THE MIDDLE EAST BEFORE 1920: POLITICAL CONTEXTS, HISTORICAL RESULTS

GEOFF EMBERLING

Although the aim of archaeological research is to acquire knowledge about the past, archaeological projects take place in particular historical and political contexts. Breasted's 1919–1920 expedition was no exception. Its success depended on Breasted's ability to make connections with political authorities across the Middle East, who gave their approval and support for his project in part because his work complemented their own.

By the end of World War I, European travelers had been acquiring artifacts from the Middle East for centuries, and Western archaeologists and historians had already been working in parts of the region for more than 100 years. Egyptian hieroglyphs were deciphered beginning in 1822 with Jean-François Champollion's reading of the Rosetta Stone, and Henry Rawlinson's publication of the trilingual rock inscription at Behistun in 1847–1849 was a breakthrough in decipherment of Mesopotamian cuneiform writing. Scholars had thus been reading texts in these scripts — as well as Hebrew, Greek, and Aramaic that had been in continuous use since antiquity — through much of the nineteenth century.[2] So the broad outlines of ancient Egyptian and Mesopotamian history were known in 1920, as summarized for example in Breasted's own popular book *Ancient Times* (1916) (fig. 2.1).

The histories of exploration and excavation, and their motivations, varied across the Middle East and through time. Biblical history and the search for religious relics drew crusaders and European travelers to the Holy Land along the eastern Mediterranean coast as early as the eleventh century, and travelers (and later archaeologists) continued to focus on biblical sites and finds in this region. The eastern Mediterranean also came to be part of the nineteenth-century travel itinerary for wealthy Europeans and Americans (as portrayed, among others, by Mark Twain in his *Innocents Abroad* of 1869). Exploration of Egypt and Mesopotamia was valued in part because of the connections of these ancient civilizations to the Bible. However, the histories and artistic traditions of the ancient Middle East were also increasingly of interest to the general public as a part of world history.

Another reason the history of the Middle East was of interest to Europeans and Americans was colonialism and the increasing inequality in international political and economic relationships. This connection between international politics and interests in research and travel can be illustrated by the simple fact that there were no Arab teams investigating the early history of America, and no Egyptian expeditions searching for the source of the Thames. It is worth considering why such projects were unthinkable.

Of course, European and American wealth and political power made it much more possible for their scholars to travel in the Middle East than for Middle Eastern scholars to travel to the West. As Edward Said, in his influential 1978 book *Orientalism*, put it, "the scientist, the scholar, the missionary, the trader, or the soldier was in, or thought about, the Orient because he *could be there*, or could think about it, with very little resistance on the Orient's part" (p. 7). But the rise of European colonial interests in the Middle East during the nineteenth century also made necessary a range of scholarly projects, including study of geography and history and ultimately, the development of the field of anthropology, which was founded in part on the efforts of colonial administrators to understand and control the varied groups living under their authority.

The colonial interests of European powers in the region included extraction of resources and access to routes to the Holy Land and to India — the overland route through Mesopotamia (Iraq), and the sea route through the Suez Canal (opened in 1869). They advanced their interests, often in competition with other European powers,

---

[2] Exploration in Anatolia (Turkey), Arabia, and Persia (Iran), with their varying histories, were not a part of the first Oriental Institute expedition and are outside the scope of this review.

PLATE I. RESTORATION OF AN EGYPTIAN VASE OF THE
PYRAMID AGE. (AFTER BORCHARD)

The original was wrought of gold (here yellow), inlaid with lapis
lazuli (here blue), by the goldsmith (§ 82 and Fig. 47)

ANCIENT TIMES
A HISTORY OF THE
EARLY WORLD

AN INTRODUCTION TO THE STUDY OF
ANCIENT HISTORY AND THE
CAREER OF EARLY MAN

BY

JAMES HENRY BREASTED, Ph.D., LL.D.

PROFESSOR OF ORIENTAL HISTORY AND EGYPTOLOGY; CHAIRMAN
OF THE DEPARTMENT OF ORIENTAL LANGUAGES
IN THE UNIVERSITY OF CHICAGO

GINN AND COMPANY

BOSTON · NEW YORK · CHICAGO · LONDON
ATLANTA · DALLAS · COLUMBUS · SAN FRANCISCO

FIGURE 2.1 Title page and frontispiece from Breasted's *Ancient Times* (1916)

through political alliances and sometimes outright political control. Even for scholars who claimed no political motivation, it is clear that their interests and work developed within a political environment and that their results could be used for political ends.

Although the Middle East is often viewed as a single unit in an oversimplified contrast between East and West, Europeans and Americans were not the only foreigners interested in the ancient Middle Eastern past. During the nineteenth century, the Ottoman Empire, which controlled most of the Middle East at that time, also began to develop an interest in archaeology with the aim of promoting the diverse and glorious pasts of the regions under its control. Osman Hamdi Bey, an orientalist painter and archaeologist, was the first director of the National Museum of Archaeology in Istanbul. Antiquities laws developed during the 1880s attempted to control excavations across the empire and led to the accumulation of an imperial collection of archaeological finds.

The history of archaeology and the ways in which the practice and results of excavation and interpretation intersect with local and international political interests have been the focus of a great deal of recent scholarly attention. The aim of such work varies from understanding of a historical moment to advocacy for change in the ways that archaeology is practiced in the modern world. In what follows, I outline the state of archaeological and historical knowledge of the ancient Middle East in 1920, with notes both on major projects and on their sources of institutional support.

EGYPT

One paradigm for the conjunction of colonial rule and scholarly knowledge of its subjects was established with Napoleon's invasion and brief occupation of Egypt in 1798–1801, to which he brought a large and well-supported team of scholars to document the conquered land, people, and their history. The magnificent result was, in part, the *Description de l'Égypte*, more than twenty massive volumes in its first edition. Arguably of equal importance was the institutional support given to the scholars and artists working on this project in the form of the Institut de l'Égypte, established in a former palace in Cairo. The discovery of the Rosetta Stone during this occupation was the expedition's single most significant find. It was later taken by British forces after Napoleon's defeat in Egypt, specifically named in the subsequent treaty between Britain and France, and is now displayed in the British Museum. This negotiation illustrates the symbolic political importance already attached to antiquities in the early nineteenth century.

In the aftermath of the French invasion, Muhammad Ali, an Albanian soldier of the Ottoman army and his successors took control of Egypt from 1805 to 1882 under the nominal control of the Ottoman Empire. Significant European involvement in the country included commercial interests in commodities like cotton as well as social causes like the abolition of the slave trade that passed through Egypt from areas to the south.

Scholarly attention on ancient Egypt during this period was focused on the removal of objects with little regard for careful excavation or documentation of existing monuments. The rush for objects was led by British and French government officials (like the British Consul Henry Salt and the French Consul Bernardino Drovetti) and representatives of their respective national museums (the British Museum and the Louvre); some of these finds also went to Turin in Italy and to the Berlin Museum. The looting of the country continued despite a series of efforts to regulate the antiquities trade, to sequentially establish a series of museums for Egyptian antiquities, and to establish an Antiquities Service (of which the early directors — Auguste Mariette, Gaston Maspero, Jacques de Morgan, and Pierre Lacau — were French). The state-sponsored Prussian expedition of Richard Lepsius (1842–1845), modeled on the Napoleonic team's work, further documented standing monuments of Egypt and Nubia. Visual portrayals of the region were not limited to scholarly studies: renderings by artists like David Roberts, who made paintings and sketches of Egypt and the Holy Land in a 1838–1839 trip, were an important part of the effort to display the ancient past to European audiences. These efforts also had the effect of separating ancient Egypt and its civilization from its modern context — monuments were portrayed as romantic ruins with only a few passing Egyptians to add scale and atmosphere to the past.

After an 1882 revolt that threatened European economic and transportation interests in Egypt, the British army invaded and established a colonial "protectorate" that lasted until 1936. Not coincidentally, it is during this period that archaeology in Egypt was professionalized and institutionally sponsored. European countries established archaeological projects that were funded by national governments (including national museums) or by public interest groups like the Egypt Exploration Fund or the German Oriental Society (Deutsche Orient-Gesellschaft). American teams began to work in Egypt, but were not funded by any government agency or national museum but rather by private sources that included wealthy individuals and major American museums that were not directly state sponsored. Antiquities laws provided for export of a division of finds, which allowed museums including the Metropolitan Museum, the Museum of Fine Arts, Boston, and the Museum of the University of Pennsylvania to gather large collections by the first decades of the twentieth century.

Among the many significant projects of this period were William Flinders Petrie's work for the Egypt Exploration Fund on Predynastic cemeteries and the early royal cemetery at Abydos, George Reisner's excavations for the Museum of Fine Arts, Boston, of Old Kingdom temples at Giza, Albert Lythgoe's excavations for the Metropolitan Museum in the Middle Kingdom settlement at Lisht, and Clarence Fisher's work for the University of Pennsylvania in the palace of the New Kingdom pharaoh Merneptah at Memphis. It was clear that part of the public interest in ancient Egypt was its connection to the Bible, yet biblically oriented research remained a relatively minor focus of these early archaeologists in Egypt.

This archaeological activity had begun to give a greater understanding of Egyptian architecture and settlement, in addition to the wealth of individual finds that could contribute to a more detailed picture of everyday life in ancient Egypt. The outlines of Egyptian dynastic history had in some sense never been lost, since the list of kings and dynasties recorded by Manetho in the third century BC was preserved, copied, translated, and commented upon. Discoveries in the nineteenth century confirmed the existence and basic ordering of the historical sequence. The problem of overlapping dynasties was not fully realized by 1920, however, resulting in dates for earlier dynasties being wrong by hundreds of years.

In the 1920s, it was common for scholars and the public to understand historical change with reference to race. Petrie, for example, proposed that a "dynastic race" had invaded the Nile Valley during Egyptian prehistory, and that they brought civilization from outside Africa. This view, widely accepted among Egyptologists until the 1950s and later, separated ancient Egyptian civilization from the rest of Africa and implicitly divided them from the modern inhabitants of Egypt.

THE LEVANT

There is no politically neutral term for the region along the eastern Mediterranean coast, extending as far inland as the Euphrates River and encompassing the modern countries of Israel, Jordan, Lebanon, and Syria as well as the Palestinian National Authority. Part of the area is called the Holy Land in both European and Arabic traditions, but the broadest term used by archaeologists is "Levant," and even that relatively obscure term is derived from the French word for "rising" (as in the direction of the rising sun, if one were standing in France).

During the nineteenth century, the Levant had been part of the Ottoman Empire for more than 200 years, and was eventually divided into a series of provinces with provincial capitals including Beirut, Aleppo, Damascus, and Jerusalem. Exploration by Europeans and Americans was not initially motivated by archaeological objects, but by locating and mapping places mentioned in the Bible and by missionary work aiming to convert Muslims or Orthodox Christians to Protestant or Catholic forms of Christianity. After several of these geographical expeditions in the earlier nineteenth century, non-governmental professional societies were formed in the United States (American Oriental Society in 1842), England (Palestine Exploration Fund in 1865), and Germany (Deutscher Palästina-Verein in 1877), and centers were established in Jerusalem (like the French École Biblique in 1890). National interests and support were also involved in several of these expeditions, including American and British attempts (1847–1848) to map the Jordan River and Dead Sea as a possible trade route from the Mediterranean to the Red Sea, and the detailed British topographical mapping surveys of biblical sites (1860s–1870s) that had obvious broader applications and was supported and funded by the British War Office.

Early archaeological excavations focused on Jerusalem beginning in the 1850s and were carried out over protests about excavations of Jewish tombs and the Haram esh-Sharif (the Temple Mount). Other early excavations included those of Ernest Renan at the Phoenician cities of Byblos, Tyre, and Sidon in what is now Lebanon in the 1860s. The labor force was composed of French soldiers, sent to protect Christians from sectarian violence in the area.

The first careful, stratigraphic excavations were not carried out until 1890, when Petrie excavated Tell el-Hesy on behalf of the Palestine Excavation Fund. This project set a new standard for archaeological excavation and also provided a ceramic chronology and evidence of cultures not mentioned in the Bible. A growing German-Ottoman friendship was marked by a visit of Kaiser Wilhelm II to Jerusalem and surrounding areas in 1898, and German excavations began soon after at sites including Baalbek and Megiddo. The first American excavation was that of George Reisner at Samaria beginning in 1909.

With tensions between the Ottoman Empire and European powers rising in the years before World War I, archaeologists for the first time used field projects as a cover for espionage. The British archaeologists C. L. Woolley (who would later become famous as the excavator of the Royal Cemetery of Ur) and T. E. Lawrence (later "Lawrence of Arabia") conducted just such a project in the Negev Desert in 1913. Woolley and Lawrence planned the work themselves as members of British military intelligence, knowing that the region would become strategically important in a war against the Ottoman Empire.

Unlike the archaeological research in Egypt, archaeology in the Holy Land during the early twentieth century was of interest less for excavated objects themselves, but for connections that places and finds might have to the Bible. At the time of Breasted's trip, biblical geography had been explored extensively, but pre-biblical (to say nothing of prehistoric) cultures were barely known. Breasted's own interests in the Levant focused on Megiddo as the site of a battle between the pharaoh Thutmose III and Canaanite armies in about 1450 BC (J. Breasted, "Battle of Kadesh").

MESOPOTAMIA

Ancient Mesopotamia — the area between the Euphrates and Tigris rivers in what is now Iraq and eastern Syria — had been controlled by the Ottoman Empire since the sixteenth century through provinces centered in Mosul, Baghdad, and Basra. During the nineteenth century, European powers began to exert diplomatic and commercial influence. British and French consuls in these provincial centers worked to facilitate the economic activities of European firms, including maintaining trade routes to India. The British East India Company in Baghdad conducted and published early explorations of the archaeological heritage of the region. In the 1830s, one of these agents (Henry Rawlinson) was able to decipher the cuneiform script by arduously copying a trilingual monumental inscription of Darius at Behistun in western Iran, nearly 200 miles from Baghdad.

The first excavations in Mesopotamia were carried out with permission from the Ottoman sultans by the French consul in Mosul, Paul-Émile Botta, who excavated in the Assyrian palaces of Nineveh and Dur-Sharrukin (modern Khorsabad) beginning in 1842. He was soon joined by Austen Henry Layard, who had been an agent of the British ambassador in Constantinople, and who began excavations at Nineveh and Kalhu (Nimrud). These projects provided tangible proof of biblical places and people. The Black Obelisk found at Nimrud, for example, depicts the Israelite king Jehu kissing the ground before the king of Assyria. A later find of a cuneiform tablet contained a written account of a great flood similar in many respects to the biblical flood. As the projects were funded by the British and French governments, the massive carved stone reliefs were sent back to the national museums — the British Museum and the Louvre — where they were received with great interest.

Explorations in southern Mesopotamia — ancient Sumer and Babylonia — developed during the 1850s, resulting in the discovery of Sumerian culture and the recovery of massive archives of cuneiform tablets that provided an increasing basis for a historical understanding of Mesopotamia that did not depend on connections to the Bible. The biblical connection remained potent, however, and the first American expedition went to Babylonia (specifically, the University of Pennsylvania at the site of Nippur) in 1888. The next American expedition would be that of the University of Chicago to the Sumerian city of Adab (modern Bismaya) in 1903–1905 (fig. 2.2).

National competition with England and France was an explicit motivation for the first German expeditions in Babylon (1899–1914) and Ashur, which were supported by the Deutsche Orient-Gesellschaft as well as by the

German Kaiser. Interestingly, the German Assyriologist Friedrich Delitzsch delivered an internationally controversial series of lectures in 1902–1904, called the "Babel-Bibel" lectures after the German terms for "Babylon" and "Bible," in which he argued against the connection between the Hebrew Bible and ancient Mesopotamian civilizations.

As in the Levant, some archaeologists worked as spies in Mesopotamia. The clearest cases of intelligence work involved excavations that also allowed for surveillance of the Berlin-Baghdad railroad that was under construction in the years just before World War I. A British team including D. G. Hogarth, C. L. Woolley, and T. E. Lawrence worked at Carchemish on the Euphrates River, now on the modern border between Syria and Turkey, from 1908 to 1911 and resumed in the spring of 1920. Baron Max von Oppenheim led a German team at Tell Halaf, 100 miles to the east, from 1911 to 1914, and was also commissioned as a spy. While the archaeological results of these projects were arguably more significant than the intelligence they produced, the projects likely would not have been started there or then without the support of intelligence agencies.

By the time Breasted reached Mesopotamia (as it was then called) in 1920, British excavators had already resumed excavation under military occupation. Their

FIGURE 2.2 Edgar J. Banks, Director of the University of Chicago Excavation at Bismaya, in native dress at the dig site, 1903(?) (OIM photograph P. 9319)

excavations at al-Ubaid used Ottoman prisoners of war as their labor force.

By 1920, scholars' knowledge of ancient Mesopotamian history was less developed than their knowledge of Egyptian civilization. Without a single list of dynasties and kings as had long been known for Egypt, chronologies and political history were unclear and discontinuous. Certainly major monuments had been discovered and the main cultures had been identified, but they were not yet connected into a coherent historical narrative.

CONCLUSION

By 1919–1920, archaeologists and historians working on the civilizations of the ancient Middle East were well on their way to understanding basic historical sequences and recognizing significant cultural developments beyond those that related solely to the Bible. It is significant that early archaeological interest in sites and monuments of Islamic times was minimal at best — Islamic history had little place within European and American narratives about their own past.

This early work also shows some of the many ways in which the practice of archaeology and the ownership of the past itself involved significant political interests. European officials were involved in acquiring objects for their national museums and also supported archaeological exploration and excavation. A few archaeologists were also spies. Private museums and societies in the United States and Europe supported excavations as a means of acquiring objects or confirming the historical reality of their own religious traditions. And inhabitants of the Levant and Egypt (at least) had begun to express their concerns about archaeological practices that included violating sacred spaces and exporting objects that increasingly came to be seen as part of local cultural heritage. The years following Breasted's trip would see the development of antiquities laws across the Middle East as well as the formation of Antiquities Departments by colonial officials as these tensions grew (see Chapter 5 in this volume for more detail).

3. THE MIDDLE EAST BREASTED ENCOUNTERED, 1919-1920

JAMES L. GELVIN

If I had any wreckage of idealistic hopes left in me when I left America, the spectacle of the Great Powers plotting against each other in the Near East has quite cured me of it. – JHB to Frances, aboard the SS Mantua, *Saturday, June 19, 1920*

When James Henry Breasted journeyed through the Middle East in 1919–1920, the region was in turmoil. Indeed, the years of World War I and those immediately following were arguably the most tumultuous in the modern history of the region. Historians cite a number of reasons for this, starting with the devastation of war. The Ottoman Empire, which had included the territories through which Breasted traveled — Egypt, Mesopotamia (Iraq), and Greater Syria (the site of present-day Syria, Lebanon, Jordan, Israel, and the Palestinian territories) — had entered the war on the side of the Central Powers, which included Germany and the Austro-Hungarian Empire, and had borne the highest proportion of casualties of any major belligerent. While estimates of German and French losses run as high as 9 and 11 percent of their populations, respectively, estimates of the Ottoman toll run from a low of about 14 percent to a high of 25 percent.

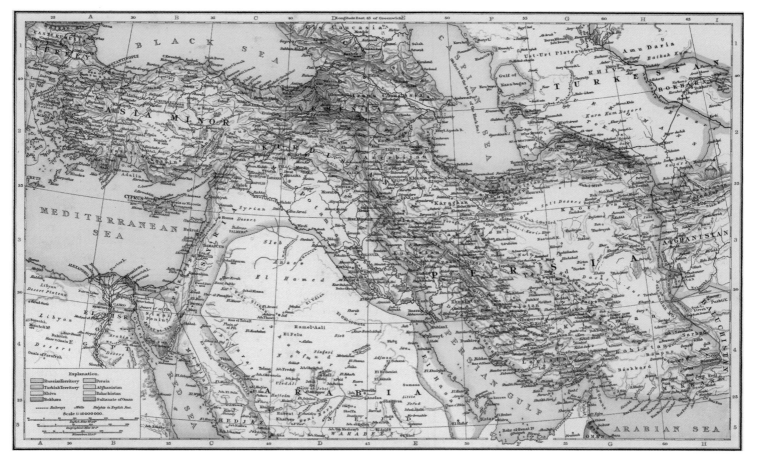

FIGURE 3.1 Map of the Middle East, 1895. From the *Times Atlas*, by Richard Andrée

Making this all the more tragic was the fact that four out of five Ottoman casualties occurred off the battle-field. The plight of the Armenians, who suffered between one and one-and-a-half million fatalities during the war, is well known. Less well known is the war's toll on the inhabitants of Mount Lebanon, where approximately 50 percent of the population died from famine induced by the entente powers' Mediterranean blockade and Ottoman inefficiencies and requisitioning policies. Lebanese talk to this very day of the *seferberlik* (literally, travel by land) — a term originally applied to Ottoman conscription, mobilization, and coastal evacuation policies but which has come to connote the entire wartime period and its horrors. Peasants in Egypt and tribesmen in Arabia also felt the effects of famine. Compounding the suffering was a breakdown of law and order throughout the empire, exacerbated by the redeployment of Ottoman troops and gendarmes from internal garrisons to frontier zones, the proliferation of armed bands whose ranks were increased by deserters, and local rebellions, some of which were induced by outside powers.

Nevertheless, the turmoil of the post-war period can only be partially attributed to the devastation caused by the war. World War I had political ramifications that unsettled the region and continue to affect it to this day. Three of these are relevant to understanding what Breasted encountered in his travels. First, there was the creation of the Middle East state system. At the beginning of the war, the Ottoman Empire had ruled, in law if not in deed, Anatolia (contemporary Turkey), Greater Syria, Mesopotamia, Egypt, parts of the Arabian Peninsula, and a small sliver of North Africa. In 1918, the Ottoman Empire was effectively dissolved. By the early 1920s, Turkey was an independent republic, France and Great Britain had divided the Asiatic Arab portions of the empire into proto-states under their authority, Egypt had evolved from an Ottoman territory to a British protectorate and finally to a quasi-independent state, and much of the Arabian Peninsula had been united under the control of the dynasty of ibn Saud (the eponymous founder of Saudi Arabia).

In addition to laying the foundations for the state system in the Middle East, the Palestine Question arose for the first time in its contemporary form as a result of wartime diplomacy and its post-war consequences. In 1917, the British government announced in the Balfour Declaration that it "views with favour" the establishment in Palestine of a Jewish home. The backing of a great power assured that the Zionist (Jewish nationalist) movement would not go the way of hundreds of other failed nationalist movements lacking so powerful a broker. It also opened the door to further Jewish immigration into Palestine and, inevitably, further confrontation with the territory's indigenous inhabitants.

Finally, with the destruction of the Ottoman Empire, no unifying political framework united Arabs and Turks — the two largest ethno-linguistic groups inhabiting the region. Nor was there a commonly accepted political framework to unite all Arabs. As a result, varieties of nationalism, often spurred on by (and indistinguishable from) anti-colonial movements, spread throughout the region as indigenous peoples, compelled by necessity, sought to define and constitute new political communities. Wherever he traveled in the region, Breasted encountered the repressive apparatus that imperial powers had put in place to check such movements.

WARTIME MACHINATIONS LEAD TO POST-WAR MACHINATIONS

Because the Ottoman Empire fought on the side of the Central Powers during World War I, Great Britain, France, and Russia — the core of the entente powers — viewed Ottoman territory as prospective spoils of war. During the war, they negotiated and signed secret agreements that arranged for the division of the empire among themselves after the end of hostilities. Each of the powers viewed these agreements as a means to safeguard its interests and enhance its strategic position in the Middle East.

France, for example, claimed "historic rights" in "Syria" — an ambiguously defined geographic unit. The French based their claim on their centuries-old relationship with Catholic and near-Catholic minorities (such as the Maronite Christians of Lebanon) who lived there and on their economic interests in the region, such as investments in railroads and silk production. The Russians covetously eyed Istanbul and its environs. Controlling the city and the Turkish Straits which it encircled would endow Russia with an all-season port and allow it direct access through the Black Sea to the Mediterranean. Russia also asserted its right as protector of worldwide Orthodoxy to ensure uninhibited access by Orthodox Christians to the holy sites in Palestine — an assertion that had led to tensions and eventually war between Russia and France during the nineteenth century. For its part, the British government had to appoint a special committee — the De Bunsen Committee — to compose a laundry list of spoils his majesty's government craved from the ruins of the Ottoman Empire. In the end, the British focused primarily on Great Britain's long-standing obsession with the security of its India colony and on ensuring post-war safeguards for British investment and trade in the region. The entente powers began their negotiations to realize their goals in 1915.

Leaving no stone unturned, the British also made wartime pledges to a number of local warlords and nationalist groups, promising to support their objectives if they allied themselves with the entente. Two of these pledges are particularly important. In 1915 the British made contact with an Arabian warlord based in Mecca, Sharif Husayn. Husayn promised to delegate his son, Amir Faysal, to launch a revolt against the Ottoman Empire to harass the empire from within — the famous "Arab Revolt," guided by the British colonel T. E. Lawrence. In exchange, the British promised Husayn gold, guns, and, once the war ended, the right to establish an Arab "state or states" with ill-defined borders in the predominantly Arab territories of the empire. The second relevant pledge made by the British during the war was the aforementioned promise to the Zionist movement. While there has been no end of speculation as to why the British would throw its support behind what was at that time a relatively small nationalist movement (British Prime Minister David Lloyd George lists at least nine reasons in his memoirs), it is most probable that the British felt that little harm could come from such a promise and that it might even prove advantageous.

Whatever written and verbal agreements were reached during the war, however, numerous factors complicated the post-war settlement:

- The various agreements and pledges made during the war were contradictory. Depending on one's reading of those agreements and pledges, for example, the entente powers had promised the territory that would become Israel/Palestine to the French, an Arab state or states, Zionists, and international control — and Palestine was not the end of it.

- Having encouraged an Arab revolt, the entente powers had to live with its result: Amir Faysal's Arab army had lodged itself in the territory of inland Syria and, with British cooperation, Faysal established a rudimentary administration based in Damascus ruling over territory on which France had designs. And, as we shall see below, Faysal was not alone among regional actors in seeking to thwart imperial designs, be they French or British.

- During the war Great Britain had launched attacks on the Ottoman Empire in Mesopotamia and along the Mediterranean coast. Although the British invasion of Mesopotamia registered initial triumph followed by disaster — British troops at first failed to take Baghdad and twenty-three thousand British and Indian soldiers died and another eight thousand were captured after attempts to rescue a besieged British army in the town of Kut al-Amara fell short — at the end of the war the British were overlords of that territory as well as Palestine, and inland Syria was under the control of Amir Faysal, whose official title was "Arab Military Commander and Advisor to the British on Arab Matters." This was the reality the French had to swallow — one which put the French at a disadvantage in negotiations.

- In 1917, the Bolsheviks overthrew the imperial Russian government and renounced (and, much to the chagrin of entente leaders and their regional allies, published the texts of) the secret agreements to which their predecessors had been party. The remaining entente powers no longer had to mollify Russian ambitions.

- Finally, America's entry into the war on the side of the entente powers, also in 1917, did little to simplify the post-war settlement. In an address to the American congress in January 1918, President Woodrow Wilson announced his famous "Fourteen Points," which he proposed as America's and the entente powers' war aims. Point I denounced "private international understandings of any kind"; Point V implied the principle of the right of peoples to national self-determination;

Point XII promised the non-Turkish portion of the Ottoman Empire "absolutely unmolested opportunity of autonomous development." French President Georges Clemenceau summed up the British and French attitudes toward Mr. Wilson's foray into international diplomacy with Gallic derision: "Even the good Lord contented Himself with only ten commandments, and we should not try to improve on them."

Thus, in the aftermath of the war, the entente powers met in Paris to unravel their conflicting claims and lay the foundations for the post-war world. At the Paris Peace Conference, the negotiators agreed to establish a League of Nations and initialed a charter for it. Article 22 of the charter dealt directly with the Middle East:

> To those colonies and territories which as a consequence of the last war have ceased to be under the sovereignty of the states which formerly governed them and which are inhabited by peoples not yet able to stand by themselves under the strenuous conditions of the modern world, there should be applied the principle that the well-being and development of such peoples form a sacred trust of civilization and that securities for the performance of that trust should be embodied in the covenant. The best method of giving practical effect to this principle should be entrusted to advanced nations who by reason of their resources, their experience, or their geographical position can best undertake this responsibility.... Certain communities formerly belonging to the Turkish empire have reached a stage of development where their existence as independent states can be provisionally recognized subject to the rendering of assistance by a mandatory until such time as they are able to stand alone, the wishes of the communities must be a principle consideration in the selection of the mandatory.

Article 22 of the charter thus applied what became known as the mandates system to the former Asiatic Arab provinces of the Ottoman Empire (the mandates system was not applied to Egypt).

The idea of mandates was new to international diplomacy and emerged as a compromise between the British and the French positions on the one hand, and that of the Americans on the other. The British and French wanted the peace conference to put its imprimatur on imperial rule over lesser-developed areas; the Americans demanded an "open door" to those areas and the abolition of imperial trade preferences, a position that was inconsistent with British and French imperial policies. While the mandatory power would have enhanced access to and influence upon its mandates, that special access and influence were to be temporary and all nations were to have equal rights in the mandates' markets.

The specific type of mandate established in the former Asiatic Arab provinces of the Ottoman Empire was known as a "Class A" mandate, and while territories in this category enjoyed what the charter referred to as "provisional" independence, they in fact occupied an intermediate space between nations deemed worthy of reaping the full benefits of the privilege and underpopulated, geographically expansive, and/or underdeveloped territories such as most of Africa and Micronesia whose independence was indefinitely postponed or denied. All nations were created equal — only, it seems, some were more equal than others. As for the right of national self-determination (a right, interestingly, first articulated by Vladimir Lenin in 1914 but never expressly enunciated by Woodrow Wilson), when it came to assigning mandatory powers, the wishes of the communities were hardly the principal consideration in the selection of mandatory — or even much of a consideration at all. In the end, Great Britain and France divided or combined territories and established them as mandates based on other principles, mainly imperial need combined with considerations of expense and practicality and projections about "the ways of the Orient" which were rooted in Christian zealotry, Romantic fantasies, and Classical and Oriental studies. Hence it was that Great Britain became the mandatory power overseeing Iraq and Palestine (later to become Iraq, Palestine, and Trans-Jordan) — entities its diplomats and soldiers fashioned and midwifed, and France became the mandatory power overseeing Lebanon and Syria — entities *its* diplomats and soldiers fashioned and midwifed.

Iraq, the mandate least prepared to face the "strenuous conditions of the modern world," was the first to receive independence in 1932, after Great Britain, finding its imperial ambitions frustrated there, effectively foreswore its obligation to the international community. One year later, Iraq's army participated in the first post-war massacre of a minority group (Christian Assyrians in northern Iraq) and two years later independent Iraq experienced its first military coup. The other mandates all had to wait until after World War II to receive their independence — although, it must be added, the future of part of the original Palestine mandate is still to be determined.

Foreign presence, continued privation, and the entente's cavalier attitude toward each community's wishes did not sit well in the region. Wherever he traveled, Breasted encountered the war's bitter aftermath. While focused on the distant past, the near past and present Breasted encountered were replete with disaffection, rebellion, and brutal repression. By the time he arrived in Egypt, the British had already suppressed a large-scale uprising, and at the very time Breasted was surveying remains of ancient civilizations in Iraq a similar uprising was brewing there. Two months before he arrived in Jerusalem, rising tensions between Arab and Jewish communities, sparked by a growing Zionist presence and assertiveness in Palestine and nourished by the institutionalized sectarianism imposed by the British, exploded into four days of intercommunal rioting. A month after Breasted's arrival in Damascus, a French army invaded inland Syria, executed or exiled nationalist leaders, and began an occupation that lasted a quarter century.

As he traveled, first through Egypt, then Iraq, then on to Syria and Lebanon and, finally, Palestine, Breasted depended upon whatever security British and French proconsuls and military officers could guarantee and, often enough, upon their hospitality. Local conditions shaped the pacing of his tour as well as his itinerary. Breasted inscribed his frustrations, observations, and concerns in letters home. These letters provide us not only with a chronicle of regional events and archaeological lore, but with a register of Western attitudes toward the Middle East and its inhabitants that Breasted shared with others of his background and station.

BREASTED'S EGYPT

> We reached Alexandria Thursday morning, Oct. 30 [1919].... There had been rioting the day before, and in fact there are disturbances of slight importance almost every day in Cairo and Alexandria.... The country people have had enough, and are quite ready to settle down under British authority; but the little tarbushed effendis in Cairo and Alexandria are still making trouble. [British Field Marshall Lord Edmund] Allenby, who was expecting to spend a long vacation in England, has already returned, and arrives this morning. There is trouble in the air, and the outbreak in Cairo is likely to come at any minute. You need not have the slightest anxiety. The trouble will be confined to certain quarters, just as was the negro rioting in Chicago. The authorities are quite ready and indeed are hoping that the lid may blow off very violently in order to show the agitators the strong hand at once and without mercy. The country is full of British troops and at Shepheard's and here (the only two hotels that are open), one sees almost nothing but khaki on the terrace and in the dining room. — *JHB to his family, Continental Hotel, Cairo, Sunday, November 2, 1919*

Although the British had occupied Egypt since 1882, when they invaded to suppress a political movement that threatened their strategic and economic interests, Egypt remained legally part of the Ottoman Empire until World War I. In December 1914, after the Ottomans had entered the war on the side of the Central Powers, Great Britain declared Egypt a protectorate, ending Ottoman sovereignty once and for all.

During the war, Egypt provided Great Britain with its largest base in the region. By war's end the British had alienated virtually all segments of the Egyptian population: large landowners could not market their cotton (the principal Egyptian export) without British controls, the educated were excluded from political power, wartime inflation impoverished civil servants and the urban poor, and famine, induced by British requisitioning of food and transport, afflicted the peasantry. The complaints of Egyptians found voice among an educated stratum of intellectuals and activists (Breasted's "little tarbushed effendis") who, at the close of war, found release from the constraints of wartime censorship and repression.

All that was needed to ignite the tensions between much of the Egyptian population and the British occupier was a spark. That spark was touched off in November 1918, when a delegation of Egyptian politicians petitioned the British High Commissioner in Cairo for permission to go to Paris to represent the Egyptian population at the peace conference. The leader of this group was Said Zaghlul, a former government minister who, while wartime vice president of the Egyptian legislative assembly, used his position to organize nationalist committees throughout Egypt.

Displaying the same tactics they so often displayed in their other dependencies like Ireland and India, the British arrested and deported Zaghlul and his colleagues to Malta. It was then that the committees founded during the war sprang into action. Demonstrations and strikes broke out throughout Egypt. They spread from students and labor activists to artisans and civil servants and even the urban poor of Cairo. Peasants, fearing imminent starvation, attacked the rail lines by which scarce food supplies might be taken to distant cities. Alongside the peasants were many rural landowners, who not only had their own complaints but who also feared social upheaval if they stood on the sidelines. The revolt (called by nationalist historians the "1919 Revolution") lasted two months before the British were able to put it down by force. Nevertheless, as Breasted's letter above attests, even a brutal show of force could not clear the atmosphere. "[I]t is going to be an exceedingly difficult task," Breasted wrote on December 10, 1919, "to restore the confidence and goodwill of the rising party of young Egyptians."

In response to the uprising, the British government appointed a commission to investigate its causes and formulate a solution. The Milner Commission concluded that Great Britain could not hope to keep direct control of Egypt and that British interests could best be maintained in Egypt if Great Britain gave Egypt conditional independence. Only then could the British hope to rein in the most vehement Egyptian nationalists. Thus, in 1922, the British granted Egypt that conditional independence. The treaty they imposed on the Egyptians was a disappointment to Egyptian nationalists. The British asserted their right to control Egyptian defense and foreign policy, protect minorities and the Suez Canal, maintain their role (alongside the Egyptians) in the governance of the Sudan to the south, and safeguard capitulations (trade concessions granted Western powers). Making conditional independence into unconditional independence would be the focus of nationalist efforts for the next three decades. Only in 1956 did the final British soldier leave Egyptian soil.

BREASTED'S IRAQ

> A few tribes refused the new arrangements. They are the kind of people for whom Mr. Wilson's 14 points are admirably suited! But Major Daley for some reason failed to apply them! He found a bombing plane more efficient. He could go out 50 miles with his pilot and bomb a tribe, come back for the usual mornings work at his desk, run out and give 'em another after lunch and transact the regular afternoon's business before tea, or postpone the bombing picnic until after tea, and return in plenty of time for a bath before dinner. The scattering of camels the first time he did this, said Daley, was very amusing. In two cases the tribal sheikh held out for fifteen days and then yielded to the discontent of his tribe and came in and submitted. If such methods are condemned on humanitarian grounds, consider the alternative…. To the Arab, "liberty" is simply the opportunity to oppress all his neighbors and raise unlimited hell. The automobile and the airplane are beginning to do what was attempted in vain for thousands of years by Babylonians, Assyrians, Persians, Macedonians, Romans and all the rest — the curbing and civilizing of the lawless Semitic or Bedwin nomads. – *JHB to Frances, Koldewey's House, Hillah, Babylon, March 30, 1920*

Saddam Hussein was not the first in the Middle East to use poison gas against his opponents. That dubious honor belongs to Great Britain's Royal Air Force, which found gas a suitable alternative to use against recalcitrant tribes when the "shock and awe" of incendiary bombs proved insufficient.

The phrase "shock and awe" is just one among many aspects of post-World War I Iraqi history that might inspire feelings of déjà vu. Others include: the overconfidence of an occupying army, which had claimed to enter Iraq to liberate it from an oppressive government; the outbreak of anti-imperialist rebellion when liberation turned into occupation, when the occupier attempted to hand-pick and install a compliant leadership to supplant accepted community leaders, and when the occupier could not guarantee law and order outside the capital; the turning of the tide against the rebellion after the occupying power mobilized and armed local collaborators who had their own complaints against the rebels; the use by political entrepreneurs of sectarian grievances and the institutional encouragement of sectarianism by an occupier who assumed sectarian identities to be immutable; the slow realization by the occupier that the game was not worth the candle and that the best it could achieve was an effective exit strategy, not guarantees of future stability; and the rekindling (and eventual betrayal?) of Kurdish dreams of national sovereignty. *Plus ça change, plus c'est la même chose.*

In the oft-repeated words of Sir John Seeley, the British had conquered their empire "in a fit of absence of mind." Nowhere is this statement as true as Iraq. The British had had their eye on the Ottoman province of Basra in southern Mesopotamia for a long time (as their interest in Persian oil and Indian security demanded), and soon after the outbreak of the war they took it. Wartime necessity explains the British advance on the province of Baghdad and the province of Mosul. But once the war ended, the British were not sure what to do with their prizes. Some in the government argued that Great Britain should rule the three provinces (by then accepted by almost everyone to comprise an entity called "Iraq") much as they did India; others argued for indirect rule. Their disagreement was settled with the establishment of the mandates system and new international norms.

Nevertheless, by the time the mandates system had been announced the British had already put in place in Iraq a civil administration modeled on that of India. British actions naturally fed suspicion of their motives, and the British refusal to allow an Iraqi delegation to present its case to the Paris Peace Conference and the arrest of prominent opposition leaders did nothing to mollify those suspicions. British proconsuls had, apparently, learned nothing from the Egypt debacle. In May 1920, Sunnis and Shias began holding joint meetings in Baghdad demanding independence. The British civil commissioner countered with a proposal for limited self-rule. If he thought this would placate anti-colonial sentiment, he was mistaken: In June 1920, rebellion broke out in the mid-Euphrates valley, then spread south and eventually encircled Baghdad, uniting Sunnis and Shias in common cause. Kurds, pressing their own demand for sovereignty, launched a separate rebellion in the north. It took the British four months to put down the rebellions (although pockets held out until 1922), which they did in large measure by convincing Sunni tribesmen that a rebel victory would mean the end of Sunni hegemony in Iraq and by experimenting with new tactics, such as the aforementioned "shock and awe." In the end, about six thousand Iraqis died in the rebellions, along with five hundred British and colonial troops.

As in the case of Egypt, rebellion convinced the British that the status quo was untenable. Confident that Great Britain's strategic interests in Iraq could be met most efficiently by appearing responsive to nationalist aspirations, the British placed Iraq on a fast track to independence, understanding full well that by leaving governance in the hands of the Sunni minority they were, in effect, guaranteeing the ruling elite's continued dependence on British friendship, if not the RAF's phosphorus and mustard-gas bombs. They even found a king for Iraq — their wartime ally, Amir Faysal, who in the meantime had run afoul of French ambitions in Syria.

BREASTED'S SYRIA (1)

> I had a very wearying day yesterday. The morning was filled with preparations for our departure to Damascus, and in the afternoon I had an appointment with General Gouraud, High Commissioner of France in Syria, at 3 o'clock, and an address before the students of the college at four. While I had only a brief conversation with Gouraud, I was impressed with him as a very strong man. When he asked me when I was expecting to come back for work in Syria, I said, "Probably not for a year, — not until all was quiet and safe". This was not wholly a diplomatic answer to give, but I could not dissemble. I mentioned the discontent of the Arabs, and Gouraud replied, "The power of France will subject them. Il faut se subir." This was a good answer for a soldier to make. His business is force. But for the French Government to set about the subjection of unwilling Syrians among whom are many educated men and who understand something of what self-government means, is no better than for Germany to undertake the subjection of the Belgians. – *JHB to Frances, Palace Hotel, Damascus, Friday evening, May 28, 1920*

Greater Syria was the one place where representatives to the Paris Peace Conference actually sought to discover "the wishes of the community" with regard to its future status. Once they did so, however, the representatives ignored what they discovered.

At the suggestion of Woodrow Wilson, the conference sent a commission of inquiry — known as the King-Crane Commission after its two leaders — to the region to solicit the opinions of its inhabitants. Although both Great Britain and France acquiesced to the idea of a commission, neither participated, and even Wilson soon lost interest. Nevertheless, the announcement of the commission stirred enormous excitement in Greater Syria. Amir Faysal, who had been in Paris representing the interests of "Asiatic Arabs" at the peace conference, returned to Damascus and called for the convening of a "Syrian General Congress" to formulate a program expressing the wishes of a majority of the population. The congress unanimously adopted a resolution demanding "absolutely complete independence" for a unified Syria — that is, the territory that includes present-day Syria, Lebanon, Israel/Palestine, and Jordan. If Syria had to have a mandatory power to advise it temporarily, the resolution declared, that power should be the United States. The second choice was Great Britain. For the delegates to the congress, France was unacceptable as a mandatory power (Breasted commented acerbically, "The Arabs did not want the French because they had to pay the French twice as much as they had formerly paid the Turks to get what they wanted, and they did not want the English because they could not get what they wanted at any price!" — *JHB to Frances, June 5, 1920*). These demands were reiterated in 1,047 (suspiciously similar) petitions submitted to the commission and in the slogan which soon became ubiquitous throughout the territory: "We demand complete independence for Syria within its natural boundaries, no protection, no tutelage, no mandate."

In the meantime, Great Britain and France worked out an understanding of their own. Having already ceded the coastal plain to the French, the British withdrew their remaining forces from the surrounding countryside and Syrian interior to Palestine, leaving the Arab-administered territory to its own devices. In March 1920, the congress proclaimed Syria independent with Faysal as its monarch. A little over a month later, entente representatives meeting in San Remo awarded France the mandate for a truncated Syria. With tensions rising between the French on the coast and nationalist guerrillas raiding into French territory, and with nationalist agitation

FIGURE 3.2 Group of Arab soldiers and civilians at Deir ez-Zor, Arab State. The lingering Ottoman influence can be seen in the *tarbush* (fez) still worn by some men. The Arab soldiers wear the very Europeanized uniforms inspired by the British. Photo by Daniel Luckenbill. May 6–7, 1920 (OIM photograph P. 7389)

escalating throughout inland Syria, the French delivered an ultimatum to the newly proclaimed Arab government. Hardly waiting for an answer, they invaded inland Syria and assumed mandatory control over what would become Lebanon and Syria.

BREASTED'S SYRIA (2)

> Our conversation then drifted to the situation in Palestine, where the position of the English seems to be steadily growing worse.... The appointment of a Jewish High Commissioner by the British politicians at home, is not a blunder of merely political consequences. It is almost certain to kindle a conflagration of the most serious proportions. Strong anti-Jewish demonstrations have already been made, many Jews have been killed and many more wounded within the last few weeks. And the commander of the British army asks, how anything else can be expected? – *JHB to Frances, Hotel Allenby, Jerusalem, Saturday, June 5, 1920*

After the French invaded inland Syria, Faysal fled, becoming in the process a British problem as well as a British ward. But this was not the only problem the British faced in Syria. Soon after the French deposed Faysal, his brother, Amir Abdallah, began advancing north from his home in Mecca to avenge Faysal's humiliation. As a result, the British now faced two problems: what to do with their wartime ally, Faysal, and what to do about Abdallah, who was threatening to make war on their more important wartime ally, France. The British persuaded Abdallah to remain in the town of Amman, which was then a small caravan stop on the route connecting Arabia with Syria, while they called a conference to determine how to deal with the worsening situation in the region. At the Cairo Conference of 1921 the British arrived at a solution. To divert Abdallah, the British divided their Palestine mandate into two parts and offered their new protégé the territory east of the Jordan River as a principality. Since this territory lies on the far side of the Jordan River (so long as your vantage point is Europe), the British called it Trans-Jordan. Abdallah made Amman his capital. No longer part of the Palestine mandate, the British closed this territory to Zionist immigration. After independence in 1946, Trans-Jordan became the Hashemite Kingdom of Jordan, which has been ruled by the descendants of Abdallah ever since. Thus it was that Winston Churchill, the British colonial secretary who presided over the Cairo Conference, would later brag that he had "created Jordan with a stroke of the pen one Sunday afternoon."

It was also at the Cairo Conference that the decision was made to place Faysal on the throne of Iraq, which he and his descendants ruled until 1958. According to a referendum organized by the British soon thereafter, that decision had the support of 96 percent of the Iraqi population — only four percentage points fewer than Saddam Hussein received in a similar referendum held in 2002.

The decision to divide the Palestine mandate into two parts and to allow or limit (depending on your point of view) Zionist immigration to points west of the Jordan River elicited angry responses in both the Zionist and Arab communities. In the Zionist community, the hardline followers of Vladimir Jabotinsky — the so-called Revisionist Zionists — never forgave the British for their perfidy and never accepted the division of the mandate. Their descendants today comprise the equally hardline Likud Party of Israel. Nationalists in the Arab community of Palestine were, at the time, mainly "Syrianists" who advocated maintaining Palestine (and, for that matter, Trans-Jordan) as part of an independent Syria (a distinct Palestinian nationalism would evolve over the years, but that is another story). They registered their opposition to Jewish immigration in a variety of forums, including the aforementioned Syrian General Congress, which resolved, "We reject the claims of the Zionists for the establishment of a Jewish commonwealth in that part of southern Syria which is known as Palestine, and we are opposed to Jewish immigration into any part of the country." For their part, the British lurched from policy to policy, and after declaring in the White Paper of 1922, "It is necessary that the Jewish community in Palestine should be able to increase its numbers by immigration," decided to limit Jewish immigration in their White Paper of 1939. The White Paper of 1939 was the last British statement on the issue before they washed their hands of the entire Palestine imbroglio. In 1947, the British abandoned their mandate and returned Palestine from whence it came — the successor organization to the League of Nations, the United Nations.

4. THE FIRST EXPEDITION OF THE ORIENTAL INSTITUTE, 1919–1920

GEOFF EMBERLING AND EMILY TEETER

PRELUDE

By 1919, James Henry Breasted had served as a professor of Egyptology at the University of Chicago for twenty-five years. He was, by all accounts, a prominent American. He was the first American to receive a PhD in Egyptology, and he was the author of important books including *A History of Egypt*, published in 1905 (which is still regarded as "probably the best general history of Egypt ever published"; Bierbriar, *Who Was Who in Egyptology*, p. 62), the five-volume *Ancient Records of Egypt* published between 1906 and 1907, and *Ancient Times*, a textbook published in 1916. His *History* and *Ancient Times* were widely read and were to prove to be important factors in Breasted's influence and success.

Through the years, as he built up the collection of the Haskell Oriental Museum (fig. 4.1), he dreamed of establishing a research institute, "a laboratory for the study of the rise and development of civilization" that would trace Western civilization to its roots in the ancient Middle East (C. Breasted, *Pioneer to the Past*, p. 238). As World

FIGURE 4.1 Photo of exhibit in the Haskell Oriental Museum, University of Chicago, ca. 1905

War I wound down, he sensed an opportunity based on the new political order, and also on his own influence. On February 16, 1919, he wrote to John D. Rockefeller Jr.:

> This whole region [the Middle East] has just been delivered from Turkish misrule, and for the first time in history the birth-lands of religion and civilization lie open to unobstructed study and research. In the entire history of knowledge this is the greatest opportunity that has ever come for the study of man and his career. — *JHB to Rockefeller, February 16, 1919*

His confidence in approaching Rockefeller, one of the wealthiest men in America, was based on Breasted's own distinguished career and the knowledge that the Rockefeller family admired his publications. In Breasted's letter to Rockefeller, he commented: "Two years ago, after reading my *Ancient Times* with your children, I believe, Mrs. Rockefeller was kind enough to write me an appreciative letter, which I prize very highly." The book would continue to be useful to Breasted.

A key point of Breasted's appeal to Rockefeller was the proposal for the establishment of what would become the Oriental Institute (see *Appendix A*). Breasted emphasized the importance of his mission by reference to scientific terms, attempting to give to the humanities the same sense of urgency and importance as the hard sciences. His Oriental Institute would be a "laboratory" where "the methods and the equipment of natural science should be applied to the study of man" (C. Breasted, *Pioneer to the Past*, p. 239). Fundamental to the implementation of his plan was a research trip through the Middle East, which Breasted had optimistically, or perhaps naively, suggested was ready to receive scholars. He suggested traversing Egypt, then reaching Basra to travel

> ... through all the leading ruined cities of Babylonia up to Baghdad; then do the same along the course of the Tigris up through Assyria to Mosul and Nineveh; and finally to cross the Syrian desert thence westward to Aleppo and the innumerable ruined city mounds of Syria. We do not expect to excavate; it is a prospecting expedition to find out <u>what</u> ought to be done under the new regime, to establish lines of cooperation with the English and French orientalists.... — *JHB to Mrs. Anderson, a donor to the Oriental Institute, December 10, 1919*

Breasted received a reply from Rockefeller (fig. 4.2) pledging $50,000 over five years for the Oriental Institute. Unbeknownst to Breasted, Rockefeller assured University of Chicago President Judson that he would pledge another $50,000 to the cause. The University of Chicago contributed additional support and in May 1919 the Oriental Institute was founded.

As outlined in Breasted's initial appeal to Rockefeller, the future activities of the Oriental Institute would be established by traveling through the Middle East. In preparation for the journey, Breasted contacted men involved with government policy in the Middle East. One of his chief goals was to express his concern for the fate of archaeological sites in the former Ottoman Empire, suggesting that a consortium of Western powers, rather than solely local Middle Eastern authorities, should have some control over them. On May 5, 1919, he wrote to William Buckler, a linguist and archaeologist who worked at the site of Sardis in western Turkey. Buckler was a member of The Inquiry, a secret task force made up of approximately 150 scholars that was convened in 1917 by Woodrow Wilson to help determine America's policy in the post-war world. The Inquiry included experts in diverse fields including Egyptology, Native American affairs, medieval history of the Middle East, and engineering. William Westermann, professor of ancient history at the University of Wisconsin, was the leader of the Western Asian Division (Oren, *Power, Faith, and Fantasy*, p. 378). The findings of the Inquiry formed the basis for President Wilson's Fourteen Points, the foundation for his foreign policy. Buckler served on a subset of The Inquiry, called the American Commission to Negotiate Peace, that was sent to Paris in 1919 to express the American vision. Breasted and Buckler corresponded about the composition of a proposed international sub-commission that would oversee archaeology in the Middle East, and their goal was to make this group a part of the provisions of the Paris peace treaties.

Breasted's desire to see Western control over archaeological sites in the Middle East was based on "the fact that the countries ... [of the Middle East] were the birth-lands of religion and civilization," hence should be shared and controlled by an international body. In his correspondence, this theme intersected with his quest for contacts and current information about the areas that the University of Chicago Expedition would traverse. He insinuated himself into the political process, reminding politicians of his deep knowledge of the ancient Middle East and how relevant that knowledge was in the current political climate. On June 23, 1919, he wrote to Charles Crane, a Chicago philanthropist and enthusiast for the Arab cause who was a good friend and strong financial supporter of Woodrow Wilson, as well as Henry King, President of Oberlin College. On the orders of President Wilson, Crane and King formed the King-Crane Commission that traveled in the Middle East to investigate Arab views of post-war politics in what was to become Syria. Characteristically, Breasted ended the letter with a glowing report of the foundation of the Oriental Institute, "a historical research laboratory," and its funding by Rockefeller. He closed with the request for any information about the "conditions of travel and the amount of money necessary to carry through the country."

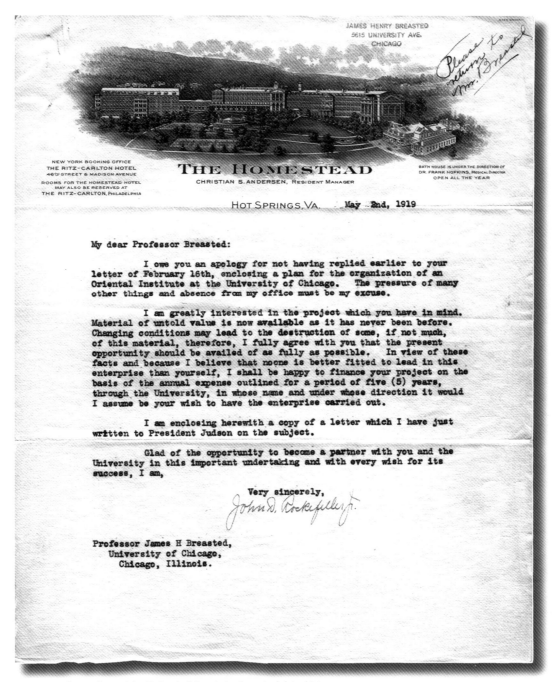

FIGURE 4.2 Letter from John D. Rockefeller Jr. to Breasted dated May 2, 1919, in which he agrees to support the formation of the Oriental Institute, stating "I believe that noone is better fitted to lead in this enterprise than yourself..."

There was a cheerful optimism and confidence inherent in Breasted's ambitious plan. America had a new prominence on the world stage as a result of its successful intervention in World War I. He may have also assumed that American presence in the area would be welcome because the United States had opted out of the mandate system, despite the discussion of being offered control over Armenia and Syria (Oren, *Power, Faith, and Fantasy*, pp. 381–82). Americans might be assured of friendly reception in the Middle East, for the provisions of President Wilson's Fourteen Points included "the principle of the right of peoples to national self-determination" and promised the non-Turkish portion of the Ottoman Empire "absolutely unmolested opportunity of autonomous development." Said Zaghlul, the leader of the Egyptian Nationalist Party, praised the American position: "No … people has felt [more] strongly the joyous emotion of the birth of the modern era which, thanks to your virile action, is soon … to spread everywhere the benefits of peace" (ibid., p. 379); and Amir Faysal, to become King of Syria and later Iraq, predicted that "Arab villagers would one day erect statues in honor of the United States (ibid., p. 383)." In his letters, Breasted referred to the "obligation" of America to undertake research in the area as a

> great new task before us and to determine the extent and character of the obligation which rests upon us Americans especially in view of the fact that we have not suffered such frightful losses in men and resources as have France and the other allied peoples. — *JHB to Mrs. Anderson, a donor to the Oriental Institute, December 10, 1919*

Breasted had two key objectives for the trip: to purchase antiquities for the museum of the new Oriental Institute and to select sites for future excavation. The group ultimately consisted of Breasted and four of his students (or former students): Ludlow Bull, William Edgerton (both graduate students in Egyptology), Daniel Luckenbill (professor of Assyriology at the University of Chicago), and William Shelton (former student who was a professor of Semitic languages at Emory University). The four met their mentor in Egypt at different times. Each of the four was expected to cover their own costs, which were considerable at approximately $2,000 each. Breasted had a different relationship with each of his traveling companions. As he wrote to his wife,

> … I presented Bull to Lady Allenby. It is a pleasure to do these things with him, for one knows that he makes a pleasant impression wherever he goes, and that a look at him at once satisfies everyone of his quality. It will not be so easy with some other members of the expedition! – *JHB to Frances, December 30, 1919*

GENERAL ITINERARY OF EXPEDITION

August 1919:	from Chicago to England, by way of New York and France
September 1919:	England
October 1919:	from England to Cairo, by way of Paris, Venice, and Alexandria
November 1919:	Egypt
December 1919:	Egypt
January 1920:	Egypt
February 1920:	from Egypt to Bombay
March 1920:	Bombay to Basra, Mesopotamia
April 1920:	Mesopotamia
May 1920:	from Mesopotamia to Arab State (today Syria) and Beirut
June 1920:	from Damascus to Jerusalem, Haifa, Cairo, and London
July 1920:	to Chicago

ENGLAND

Breasted arrived in England on August 29, 1919. For the next six weeks, he busily made arrangements for his onward travel. On September 1, he visited the British Military Permit Office to obtain the necessary permission to travel to Egypt and Mesopotamia. The office was "swarming with people." After waiting for hours, at five o'clock, although the office was still filled with people seeking their permissions, the officials declared the office was closing for the day. Again, Breasted's publications and his initial political connections led to success:

> I stepped up at once and dropped my card on his desk, and said, "Excuse me, Sir, but I have letters from the British Embassy." He was looking at my card, and before I could go any further, he said, "Why, I know you, Sir, I have read your books and I owe you a great deal. What do you want me to do?" I had been warned by Worrell to be on my taps — that this was a hard man to get past; but all was now easy. We had a delightful conversation. His day was ended at his desk, and he walked with me to my hotel, where he came in and had some tea. I went up and got him a copy of *Ancient Times* and wrote his name in it. He was greatly pleased.... This morning he put me through and gave me every thing I asked him, and then went over to the French officials, who also have a say, and put in my papers at the head of a long line of waiting people, and finished for me in five minutes, what would have taken me probably the whole morning. – *JHB to his family, September 2, 1919*

Ancient Times proved essential in another situation with the Foreign Office. On November 2, 1919, Breasted wrote a letter home, explaining that because he was "without a friend to advise me or to push my cause in the Foreign Office," he sent a copy of the book to Mr. Balfour, the British Foreign Minister. In the attached note he flattered the recipient, referring

> incidentally to his own writings, [and] told him what I was planning to do, and simply said that his influence would be quite decisive in settling whether we got out to Babylonia or not, and I hoped that I might count upon it. There was not time to receive an answer in London, so I asked him to write me in Cairo. Now such matters are often referred to subordinates in a great ministerial office and hang fire for weeks before a reply is sent; but on my arrival here [only days afterward?] I found the following letter from Mr. Balfour:
>
> "Dear Professor Breasted:
>
> "Let me first thank you for your admirable history, which I received with the very greatest pleasure. As regards your archaeological expedition, I shall be very glad to do anything I can. I am not at the moment administering the Foreign Office; but I have taken the liberty of sending your letter to the authorities there with a request that they might do their best to make all the necessary arrangements.
>
> Yours very sincerely,
>
> (signed) Arthur James Balfour"
>
> [October 23, 1919]
>
> "With reference to your letter of the 6th October, addressed to the Right Honorable A. J. Balfour which has been passed to the Foreign Office, I am directed by Earl Curzon of Kedleston to inform you that a letter has been sent to his Majesty's High Commissioner for Egypt and the Soudan requesting him to accord you every assistance in his power as regards your proposed journey from Egypt to Bosra.
>
> I am, Sir, Your most obedient, humble Servant
>
> (signed) T. Wellesley"

Breasted's publications yet again played a key role in establishing valuable political connections. Taking advantage of High Commissioner Allenby's presence in Britain, Breasted wrote:

London is making great preparations to receive him. I have taken the bull by the horns and written him a letter stating exactly what I must have if our expedition is to get out on the ground at all. I have sent him a copy of my *History of Egypt* which, as I had to procure it here is not as pleasing as the American edition — no gold except on the back. I addressed both to the War Office, and now I shall have to hang around and see if anything happens. – *JHB to Frances, September 15, 1919*

The ploy worked. Breasted received copies of two letters from Allenby, one to his Chief of Staff, and the other to the Acting High Commissioner Sir Milne Cheetham "who rules Egypt in Allenby's absence, — asking them both to do 'everything possible' for me. I think that will solve my difficulties" (*JHB to Frances, October 1, 1919*). In awe at being close to men of great power, Breasted related that the letters had been written while Allenby was visiting the king at Balmoral Castle. In another letter, Breasted triumphantly wrote:

The High Commissioner, as you know is Lord Allenby, who has already done everything I have asked, without any official backing. He has now been instructed by his government to do every-thing he can for me, and with Mr. Gary's assistance also ensured, I have no anxiety about our transportation. But it has taken lots of nerve for a backwoods boy from the Illinois prairies and no end of effrontery in acting as if I had "always come down stairs that way." – *JHB to his family, November 2, 1919*

The hours that Breasted spent pursuing permissions for his travels in Egypt and onward made him more of a realist about the political situation of the area. He warned his Chicago colleague George Allen: "I should have to write volumes to describe the situation here. The discouraging aspect of it is that even in archaeology, it reeks with politics, and intrigue and counter-intrigue are everywhere" (*JHB to T. George Allen, October 2, 1919*).

Not only was Breasted becoming more attuned to political intrigues, but he also was gaining valuable insight into the intersection of politics and antiquities. On his visit to Lord Carnarvon, the prominent collector and pa-tron of Egyptian archaeology, Breasted recorded that the conversation "... was almost wholly devoted to a careful discussion of the international politics involved in archaeological work in the Near East." Carnarvon gave him "valuable letters to British officials in the Orient" (*JHB to T. George Allen, October 2, 1919*) and "volunteered to write me letters of introduction to the commander in Mesopotamia and several other leading men in the British administration in Asia. That means all sorts of help, like motors, transportation, and even airplanes if I ask for them. So the visit was a valuable one in a number of ways" (*JHB to Frances, September 15, 1919*). He also became aware of the parallels between the political rivalry between the English and French for domination of post-war Middle East and their control over antiquities in Egypt, a diplomatic battle that had been ongoing for more than a century. In contrast to his support for the British, Breasted was critical of the French official who ran the Egyptian Antiquities Service and the restrictions that they increasingly placed on the exportation of antiquities. Some of this criticism was clearly the result of Breasted's own view of the Egyptians, who he felt had little interest in their own past:

Lacau, the French Director General in Egypt, succeeding Maspero, is a very sincere and upright Frenchman, and a fine scholar; but he is very unsuccessful as an administrator, and is an idealistic dreamer. He believes in Egypt for the Egyptians to such an extent that he is now definitely plan-ning for the abandonment of the old policy of a fair division with the foreign museums carrying on excavations, and wishes to swamp the Cairo Museum, already far too large, <u>ten times</u> too large for the administrative staff, with an inundation of monuments which it cannot possibly install or administer. Meantime he forgets that the number of educated Egyptians who can appreciate such things is an insignificant handful, while on the other hand, as our birthright and inheritance from the past, Egypt can be a wonderful educational influence in civilized lands of the West by means of the remarkable collections which it can furnish without in the least injuring the Egyptians of today or the Cairo Museum. – *JHB to Charles, September 25, 1919*

Armed with letters from Allenby's office, Balfour, and Wellesley, Breasted crossed the channel for France at the end of the first week of October 1919.

FRANCE

Breasted's increasing awareness of the aftermath of World War I gained from conversations with British officials took on a more concrete form in France. He took a package tour of the French battlefields, recounting, "I remember following these operations on a map at the time, only a year ago last July" (*JHB to his family, October 11, 1919*). He toured a now abandoned dressing station "still equipped with rough beds on which hundreds of dead and dying men had lain, as the over-worked surgeons endeavored to reach them." At Reims, he was confronted with the sight of German prisoners restoring the ruined Reims cathedral. It is possible that these experiences gave him a greater appreciation of the European powers' desire for tangible territorial gain in the Middle East in light of their tremendous losses.

Breasted's eight days in Paris were devoted to establishing contacts with academic peers rather than political contacts. He also had a busy and fruitful schedule of visiting dealers and studying the collections at the Louvre. During that week, Breasted purchased over 700 objects for the University of Chicago. He commented on how intense and exhausting it was to go through the dealers' offerings. He visited the home of the Kalebdjian Brothers who also conducted business in Cairo:

> They also had an entire house filled with wonderful things which they were offering for sale.... I went through their entire stock, which was a job of days, like going through a considerable museum, piece by piece, — slow and wearying work. I usually kept going until 7:30 or even later, and then went off to dinner and bed. – *JHB to his family, October 18, 1919*

Breasted selected most objects for the Oriental Institute with an eye for their instructive value. In some cases, he knew more about the objects than the dealers did. After purchasing a large selection of colorful wall inlays (fig. 4.3) from a thirteenth-century BC palace in the eastern Delta from the stock of Kalebdjian Brothers, he wrote:

> There is a great deal of fragmentary material from Tell el-Juhudieh, representing the beautiful glazed incrustation of the palace. The dealers had not noticed that a number of the pieces fit together, and we shall be able to build up the designs, at least partially. – *JHB to Frances, November 3, 1919*

From the same dealers, he purchased a rare example of a water clock (fig. 4.4) showing the god Thoth in the form of an ape.

On October 17, Breasted took the famed Orient Express train to Venice where, after a brief rest, he boarded a ship for the eight-day voyage to Alexandria.

EGYPT

Once in Egypt, Breasted began to use his connections. Upon landing in Alexandria, he found that

> ... to our disgust the captain could not find room at the docks and told us he would not be able to dock until 4 P.M.! So I showed my letter from Lord Allenby to the port authorities who were examining passports and the officer in charge at once said he would give me a pass to take a felucca for the dock three miles up the harbor.... As I entered the customs there was the familiar room where I had cleared our trunks thirteen years ago, when the little French doctor got my red bag of silver! My letter from Allenby was magical. The chief of the douana merely asked me to give him my card bearing an assurance that I had nothing dutiable or contraband, and passed the whole lot without opening a thing; while the Clays[3] had to open everything to the last nook and corner. – *JHB to his family, November 2, 1919*

---

[3] Albert T. Clay was an Assyriologist from Yale who was traveling to Jerusalem for a one-year teaching appointment at the American School. Breasted disliked Clay, initially for the fact that Clay brought his wife and daughter with him "when the difficulties of getting transport for food are so great." Breasted also criticized him for his hygiene, dress, and overbearing attitude.

FIGURE 4.3 Group of inlays from the palace of Ramesses II at Tell el-Yahudiya in the eastern Nile Delta, purchased in Paris from the Kalebdjian Brothers. Breasted bought more than 200 of the rosettes in different sizes. Dynasty 20, ca. 1184–1153 BC (OIM 9864B, 9864D, 9868C, 9866B)

FIGURE 4.4 Model of a water clock in the shape of the god Thoth as an ape. Dynasties 26–31, 664–332 BC (OIM 10101)

The political situation in Cairo was tense. He wrote about unrest in the streets of Cairo and Alexandria and dismissively referred to the political agitators:

> But these troubles seem to be confined to these two big towns. The country people have had enough, and are quite ready to settle down under British authority; but the little tarbushed effendis in Cairo and Alexandria are still making trouble. Allenby, who was expecting to spend a long vacation in England, has already returned, and arrives this morning. There is trouble in the air, and the outbreak in Cairo is likely to come at any minute. — *JHB to his family, November 2, 1919*

In another letter, he recounted the level of tension and how censorship had kept the violence from the public:

> The English have a very difficult situation on their hands here. Preoccupied with the pressing affairs of the Paris Peace Conference, Lloyd George did not give the legitimate claims of the Egyptians, which the English have never intended to disregard, the immediate attention which they ought to have received. And it is going to be an exceedingly difficult task to restore the confidence and good will of the rising party of young Egyptians. There has been almost daily violence in the streets of Alexandria, and frequently here in Cairo, sometimes with loss of lives. The revolutionaries have now taken to assassination, stealing up behind solitary British officers after sundown and shooting them in the back, sometimes even when they are walking with a lady. Of course these things are censored out of the dispatches you see in our papers. — *JHB to Mrs. Anderson, December 10, 1919*

As he talked with officials, he gained keener perspective on how widespread the unrest in the Middle East was. Mr. Greg, Head of the British Foreign Office in Cairo, shared a report with Breasted:

> "That is confidential, and I must ask you to say nothing about it, but it is important and you ought to know it." My eye fell on a big rubber stamp marked "SECRET," then on the heading: "Armée Française en Syrie," and I found myself presently deep in a report from French headquarters.... It is evident that the whole middle section of the Fertile Crescent from Baghdad to Aleppo and Damascus is on fire, and a concerted effort is being made by the Turks and the Arabs to throw the French into the sea. We shall not get far from Baghdad, I fear. Be quite free from all anxiety. We shall run no risks, and shall turn back and come quietly home by the route we came over, whenever it seems hazardous to go any further. — *JHB to Frances, February 18, 1920*

Allenby's return to Egypt gave Breasted another view of the nature of the British occupation of Egypt:

> As the train pulled in the entire air force, a beautiful squadron in V formation, swept over from Gizeh, circled over the station and followed the High Commissioner's carriage from the station to the Residency. It was an imposing demonstration of British power in the East, and it must have given the hysterical nationalists here something of a jolt. — *JHB to his family, November 11, 1920*

Through his connections, Breasted became a sort of a political insider and he was invited to a variety of social gatherings with British military officials. He recalled the setting of a ball and his conversation there with Major General Bols, Lord Allenby's Chief of Staff:

> It was like a scene from one of Mrs. Humphry Ward's novels, this brilliant ball room, filled with the big men of the British Empire who are out on the frontiers doing things, and taking their relaxation in a roomful of beautifully dressed and pretty women, and doing it with great gusto and evident enjoyment; while all around the air is keen with rumors of impending trouble. It was indeed a fascinating experience to stand in a corner with one of the leading men in the situation and watch the whirl of the American dances, which after all we scarcely saw, as we talked of the big game of modern empire in the Near East and the grave dangers which French insistence on coming in and taking Syria, has introduced into the situation. It means the continuance of the world-war, and the French army, or its best men, are as much aware of the fact, as their Foreign Office is ignorant of it. — *JHB to his family, November 10, 1919*

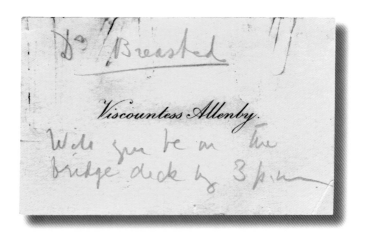

FIGURE 4.5 Calling card of Viscountess Allenby left for Breasted with the notation "Dr. Breasted, will you be on the bridge deck by 3 pm"

He was invited to dine with Lord and Lady Allenby (fig. 4.5). This opportunity gave him the chance to have his first conversation with the Lord High Commissioner. He presented himself in the role of potential diplomat and as a possible source of useful political information:

When he entered the drawing room before dinner he had greeted me very quietly but with a kind reference to the correspondence we had had. My History of Egypt lay on the table in the library, and I had handed him a copy of my study on the Battle of Kadesh as we went out to dinner. After dinner as the gentlemen drifted into the drawing room following the ladies, to my great surprise Allenby dropped into the group at my side, led me to a chair apart from the company, and seating himself, began to take up a remark I had made referring to Clemenceau's <u>bon mot</u> "Le bon Wilson avec ses quatorze points et le bon Dieu qui n'en a que dix."[4] From the beginning of the evening in the drawing room, when Allenby thus seated himself, we sat with our heads together, and he continued to talk without interruption or addressing a single word to his other guests, until the company was broken up by Lady Clayton's rising to go.

Whether this was due to a remark of mine expressing a desire to know all I could of the present situation in the Near Orient, I do not know; but I had taken the first opportunity to say that I hoped to further the establishment of cordial cooperation between his country and mine in the future control of the Near East, and for that reason I would be glad of an opportunity to learn all the facts regarding the situation which it might be proper for me to know. He made no response whatever, but he at once began to talk. — *JHB to Frances, November 30, 1919*

The subject of Megiddo came up when Allenby referred to his title "Lord of Megiddo" which was awarded on the basis of his victory there over the Turkish army. This gave Breasted the perfect entrée to demonstrate his knowledge of the past and how relevant ancient studies were to the present situation. Breasted recorded Allenby's comments: "Curious, wasn't it, that we [British forces] should have had exactly old Thutmose's experience in meeting an outpost of the enemy and disposing of them at the top of the pass leading to Megiddo! You see, I had been reading your book and [George] Adam Smith, and I knew what had taken place there" (*JHB to Frances, November 30, 1919*).

Breasted tried to make himself as useful as possible to the British military, stressing how a scholar of the past could be of assistance in the current political situation. Shortly before he left Cairo, he met three times with members of the Milner Commission whose task was to negotiate an alliance that would recognize Britain's special position in Egypt after the country's independence. He wrote:

They are a very fine group of Britons, but they are confronted with an insoluble problem and an impossible task. I think they know it, too. Lord Milner was very kind, but seemed more interested in talking my shop than his, at which of course he is grinding all day long and every day. One of the tasks I have left unfulfilled was to write a letter for the use of the Commission on the state of the Antiquities Department and what ought to be done.... I want very much to do it, for the condition of the Antiquities Administration is lamentable and the loss to science and the world is incalculable. You know Lacau is now in Maspero's old post as head of the Department, and it is the case of a good scholar put into an administrative post which he is unable to fill. — *JHB to Frances, February 18, 1920*

---

[4] "Wilson with his Fourteen Points, and the good Lord who only has ten."

Breasted's association with Allenby and his staff ultimately ensured the ability of his group to carry on from Egypt, for Allenby gave him yet another letter authorizing and ensuring passage from Bombay to Baghdad. He also received a letter from Lord Carnarvon's associate Major General Percy Hambro who was in command of logistics in Baghdad, and to whom Breasted, predictably, had sent a copy of *Ancient Times*. Hambro promised his help and support, and even supplied advice on hotels and transport in Bombay and Basra and authorized Breasted to purchase supplies from British Commissary stores, the last favor being a welcome help to the expedition's budget.

One of Breasted's specific requests was a letter of introduction to Faysal, then king of the Arab State, whom Breasted described as "the only man in Asia who could protect us among the Arabs, if we were foolish enough to go among them." He obtained the letter from Allenby himself.

In addition to making political contacts, Breasted's fourteen weeks in Egypt were filled with purchasing antiquities for the University and the Art Institute of Chicago. He recognized that the war years had made it difficult for the dealers to sell antiquities. As he had written to University of Chicago President Judson:

> ... I again urge the importance of the unprecedented opportunity resulting from the fact that the usual volume of travel in the Near East has been suspended for nearly five years. This condition of affairs has left the oriental antiquity dealers and other natives without any means of disposing of the valuable things which have come into their hands season after season for five years. The opportunity awaits the first comers, and our new Oriental Institute permits us to be among these; but its budget is not large enough to meet the emergency and furnish us with a purchasing fund. – *JHB to Judson, August 7, 1919*

He exhibited his ability to make exciting and meaningful connections between the past and present, appealing to Judson's practical business side for funding:

> We want to be able to buy a considerable body of the ledgers and day-books, notes and leases, contracts and business accounts of the ancient merchants of Western Asia, whose daily records of business, written on clay tablets, beginning over five thousand years ago, furnish us with the origins of the very business forms and customs which make up the daily practice of business at the present time. – *JHB to Judson, August 7, 1919*

In addition to the initial $18,500 from the University and $500 from James Robinson of Inland Steel, he carried a letter of credit for $5,000 to purchase for the Art Institute of Chicago.[5] Yet a recurring lament in his correspondence with Judson was how much more could be done if he had a larger budget.

At this time, buying and selling antiquities in Egypt was legal, indeed, the Egyptian Museum itself had a "Salle d'vente" (Sales Hall) where duplicate or unwanted statues, coffins, and reliefs were sold. The major dealers that he purchased from included Panayotis Kyticas, Nicolas Tano, Maurice Nahman, André Bircher, Ralph Blanchard, and the Kalebdjian Brothers. Breasted wrote letters full of colorful details about the dealers and their "canny" ways. He described the home of André Bircher in Cairo in a letter to his family:

> It is an ancient house built some 450 years ago, with wonderful old Saracen carving and antique glass in the open work of the fretted stone windows. Here Anton [sic] Bircher has lived for nearly fifty years, conducting a little office just off the spacious court below, and carrying on there an importation business in which he has amassed a fortune. For nearly forty years he has been buying antiquities and he has an immense mass of stuff. He has an elderly woman as curator to look after it all, and after serving us oriental coffee under the afternoon light coming through the wonderful ancient glass and shimmering over a fountain in marble mosaic in the floor, he left us to go back to the office where he has spent half a century, and the lady took us around the collection. Nine tenths of the stuff is junk. Of the other tenth, he has sold off much that was valuable. – *JHB to Frances, December 14, 1919*

---

[5] Adjusted for inflation, the sums in 2009 dollars are $235,500, $6,370, and $63,700, respectively.

He threw himself into the task of visiting the dealers:

> I am trying to do the work of three men at least and perhaps more. There are first the antiquities to be purchased for the museum at Chicago. I spend hours a day looking over the materials here in the hands of dealers. It is endless; each stock like a museum which has to be gone over. This afternoon I began going over the cellar magazines of the great Cairo Museum where there are vast masses of things doing nobody any good, and which I am trying to secure for Haskell. I mussed through the dust and filth of a small fraction of it only. I must also spend as much of the day as I can on the museum collections copying unpublished inscriptions. – *JHB to his family, November 10, 1919*

Among Breasted's greatest purchases was what is now known as Papyrus Milbank, a Ptolemaic Book of the Dead (fig. 4.6A–B). Buying a fine papyrus (actually two, as it transpired) was a stated objective. In August 1919, he wrote to Martin Ryerson, President of the Board of Trustees of the University of Chicago:

> Unfortunately, there is not a single old Egyptian papyrus in the Haskell Oriental Museum [the forerunner of the Oriental Institute], except a few tattered fragments of the Book of the Dead. We greatly need a body of papyri.... These are constantly appearing in the hands of dealers along the Nile.

Breasted recalled that Nicolas Tano (whom he described as a "particularly hard-headed Greek") mentioned he had a special papyrus for him, but he discounted it as a "cock-and-bull" story.

> So I went over with Tano, for the place was just across the street, and after some parlaying, he secured possession of a mysterious box which we brought back to his own shop. I thought of the ragged and tattered masses of papyrus which I had handled at Nahman's — the kind of thing we always think of when we hear of papyri just out of the ground. For in almost all cases they survive as worm-eaten fragments, rarely showing any resemblance to a roll. If a roll does survive, the natives who find it usually break the roll straight across as one would break a stick, in order to divide the plunder. So after Tano had carefully locked his shop door, I was only moderately interested as he began to open his box. When the lid came off, I saw a lot of mummy cloth bandages lying under it and said to myself, "Of course it is the usual mess of tatters." And then I could hardly believe my eyes, for I saw something which I have never yet seen in all my years in Egypt. Tano lifted the

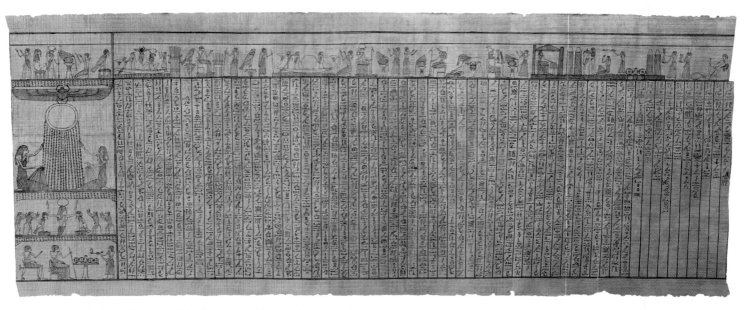

FIGURE 4.6A Section of Papyrus Milbank, a Book of the Dead, purchased from Nicolas Tano in Cairo. Ptolemaic Period, fourth–third century BC (OIM 10486B)

FIGURE 4.6B Detail of Papyrus Milbank

mummy wrappings, and lying under them was a beautiful brown roll of papyrus, as fresh and uninjured as if it had been a roll of wall paper just arrived from the shop! And it was about as thick as an ordinary roll of wall paper!... I confess I had some difficulty in maintaining a "poker face." ... Tano laid it down on the table, put his finger on the unrolled inch or two and giving the roll a fillip, he sent it gliding across the table, exposing a perfectly intact bare surface before the beginning of the writing. It was the first uninjured beginning of a papyrus I had ever seen unrolled and the first roll I ever saw in such perfect condition that it could be thus unrolled as its owner might have done. And then came the writing! An exquisitely written hieroglyphic copy of the Book of the Dead with wonderfully wrought vignettes, the finest copy of the Book of the Dead which has left Egypt for many years! – *JHB to Mrs. Anderson, December 10, 1919*

On a later visit to Tano's home in Heliopolis (fig. 4.7), Breasted, accompanied by Ludlow Bull, made a major purchase — a set of Old Kingdom statues which are now among the highlights of the Oriental Institute's Egyptian collection (figs. 4.8–9). There, Tano

showed us with much secrecy a group of 25 remarkable statues of limestone.... Four of them depict a deceased noble and his wife; the others his servants and members of his family engaged in all sorts of occupations for his comfort and enjoyment. Three of them are playing the harp, one is slaughtering and quartering a beef, a group are grinding flour, mixing and kneading dough, molding loaves and baking bread; others are cooking food over a fire, one is mixing beer and another decanting and sealing it in jars; one is turning pots on a potter's wheel. They are all colored in the hues of life... they form together as one sees them arrayed on a large dining table, a bright and animated group like a picture out

FIGURE 4.7 View of the Cairo neighborhood named Heliopolis, from the roof of Villa Mandofia where Breasted stayed (OIM photograph P. 25966)

FIGURE 4.8 View of the group of serving statues at the home of Nicholas Tano. Edgerton, whose face is just visible to the left, is holding up a fabric backdrop for photography. February 5, 1920 (OIM photograph P. 7775)

FIGURE 4.9 Statue of a man butchering a calf, from the group of serving statues purchased from Nicolas Tano in Cairo. Dynasty 5, ca. 2445-2414 BC (OIM 10626)

FIGURE 4.10 Worked flints from the collection of Captain Timins, purchased by Breasted in Cairo from Panayotis Kyticas. Dynasties 3-4, ca. 2686-2498 BC (OIM 11211, 11219)

of the real life of nearly 5000 years ago, when Europe was still in the Stone Age, and the cultivated life of Egypt was already possessed of highly developed arts, and its society had already produced sculptors who could put such life into vivacious groups in stone. – *JHB to Frances, January 25, 1920*

In some instances, Breasted was offered whole collections of objects. A Captain Timins who served in the British army in Egypt had assembled a group of over fifty very fine worked flints (fig. 4.10). Timins offered the collection to Breasted. He wanted to purchase it, but he hesitated, hoping to get a more favorable price. Only days later, when visiting the dealer Panayotis Kyticas, he was offered the same collection which, to Breasted's dismay, Timins had sold in the meantime. Breasted could not pass up this valuable material a second time even though he had to pay the dealer more than Timins had initially asked for. As Breasted wrote "I have lost 50 pounds, – a most vexatious malheur when you are trying to make your funds go just as far as they can possibly be stretched" (*JHB to Frances, December 5, 1919*).

Agents from other museums as well as private collectors were making the rounds of the dealers in Cairo, often resulting in competition for objects. On some occasions, Breasted returned to a dealer to finalize a purchase only to find that it had been sold to another buyer. He often had to make quick decisions:

To give you an idea how fast things go, I was just leaving Tano yesterday to go to lunch, when he said that he had a collection of cuneiform tablets [fig. 4.11] which a merchant of Aleppo had brought over with him. They had been examined by Sayce, who had arranged for the Dublin University to buy them; but the arrangements offered by the University were unsatisfactory and I could have them if I wanted them. Lunch was at once forgotten; Tano brought out a box, filled with neatly packed little

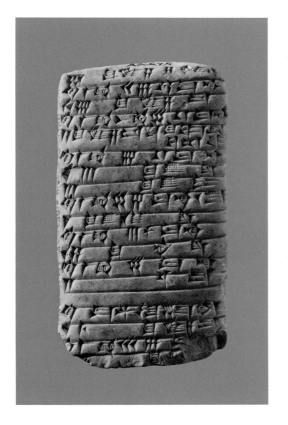

FIGURE 4.11 Selection of cuneiform tablets purchased by Breasted from Nicolas Tano in Cairo, 1919. Economic texts from the Ur III period, ca. 2100–2000 BC (clockwise from left: OIM A2645, A2651, A2638, A2655)

FIGURE 4.12 Figure of the god Amun purchased from Maurice Nahman in Cairo. The base is inscribed with an important text of King Amasis, naming his daughter Nitocris as the First Priest of Amun. Dynasty 26, 570–526 BC (OIM 10584)

packets, each containing a beautifully preserved cuneiform tablet. There were letters, contracts and things I was not sure about.... There were 258 of these tablets.... – *JHB to Frances, November 23, 1919*

Breasted was pursued by the dealers who recognized that he was a serious customer. He described the purchase of a group of bronze statues from Maurice Nahman, including an important figure of the god Amun inscribed for princess Nitocris II (fig. 4.12): "I have secured some wonderful things. A noble collection of bronzes selected from a whole series of many hundreds which have been collecting in the hands of dealers during years of the war" (*JHB to Frances, January 25, 1920*).

Shortly before he left Egypt, Breasted traveled south to Luxor. There, he and Bull met their fellow travelers William Shelton, William Edgerton, and Daniel Luckenbill. Breasted, in some cases with his fellows in tow, made the rounds of the Luxor dealers. Just before the group left Egypt, Breasted's entreaties to Judson for an additional $5,000 were met with a cable advising him that the University was sending another $25,000. Breasted's response was: "Something of a Christmas present!" (*JHB to Frances, December 24, 1919*). As he wrote: "I was rushing about among the dealers at a desperate pace, endeavoring to rescue the fine things which my new funds enabled me to secure" (*JHB to Frances, January 25, 1920*). His purchases included "an XVIIIth dynasty officer's battle axe which he carried in the fifteenth century B.C., with bronze

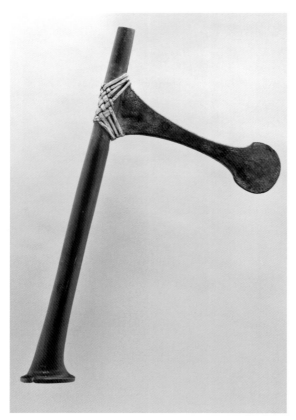

FIGURE 4.14 Far right, the antiquities dealer Yussuf Hasan. The others, left to right, are Ludlow Bull, Mrs. J. Collier, Miss Huxley, and Mr. J. Collier. Breasted took the photo. Luxor, January 1920 (OIM photograph P. 6722)

FIGURE 4.13 Bronze battle axe purchased in Luxor from Mohammed Tadrous, January 1920. The wood haft and leather lashings are original. Dynasty 18, ca. 1500 BC (OIM 10548)

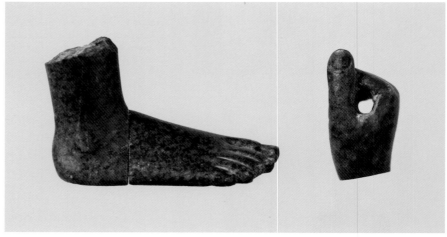

FIGURE 4.15 Lapis hand and foot from a small-scale composite statue purchased from Yussuf Hasan in Luxor, 1920. Dynasties 22–26(?), ca. 945–525 BC (OIM 10517A, B)

head, fine wooden handle and leather lashings all in perfect condition." (*JHB to Frances, January 25, 1920*) (fig. 4.13). Among his other important purchases is the fine cartonnage mummy case of the Temple Singer Meresamun from Yussuf Hasan (fig. 4.14). In the same letter to Frances, Breasted wrote of his meeting with that dealer:

> He rummaged in a crazy old safe built into the thickness of the walls of his house and brought out one treasure after another. Among them was a lovely little hand, and likewise a foot (fig. 4.15) carved with marvelous refinement in deep blue lapis lazuli, — part of a wondrous statuette wrought by some forgotten master living at the imperial court in this great capital of the East when Egypt was ruling the whole eastern Mediterranean world. And so I could go on indefinitely. I secured these things for the University, and many others.

As Breasted and his colleagues made the rounds of the dealers, they kept an eye open for material that would provide appropriate dissertation materials. As related in a letter to his wife, Breasted expected students, if able, to pay for the subject of their doctoral research!

> I have found a large massive rectangular cedar coffin at Bircher's with texts from the Middle Kingdom written on the inside, — what I call Coffin Texts in my Morse lectures. They will make a fine body of material for Bull's dissertation. Bircher had sold it to Brussels before the war, for 400

pounds Egyptian; but that is now off and he will take 350 from Bull, who I think will buy it. I am very much pleased to have settled Bull's dissertation work in this way. The Assyriologists have no difficulty in getting new documents for their students, for clay tablets are cheap and plentiful; but it is quite difficult for us to find new material of this kind for our students. I hope to do something similar for Edgerton, who unfortunately is very late in arriving, and will not reach here much in advance of Luckenbill; but of course Edgerton cannot spend any money and it must be something the university can properly buy. – *JHB to Frances, December 30, 1919*

Breasted spent a considerable amount of time purchasing artifacts for the Art Institute of Chicago. He had been an advisor to them since 1896 when he catalogued part of their collection, and he kept in close contact with their board and President Charles Hutchinson,[6] later writing:

> I think that I have never been so busy before in all my life and I have seen some fairly busy times. I have spent a great deal of time on the Art Institute purchases, and it has been a great pleasure to do so, for I have secured you some very beautiful things. I have been through the entire stocks of the leading dealers in Cairo; chiefly Blanchard, Kyticas, Tano, Nahman and Kelekian.... It has taken a great amount of time to go over these collections.... I feel however that I am in duty bound to let you know that the present opportunity to secure more such material will never return again, and that it would be very wise to seize the opportunity while it is still ours. The situation is this. The natives have made a great deal of money on the war. Many of them who never bought antiquities before have done so since last spring and they are holding all that they have bought at preposterous prices.... Meantime most of them are willing to listen to reason and are disposing of what they have at practically pre-war prices ... there is therefore a body of material here in Cairo, which will never be available again and which would give the Art Institute at fair prices a very beautiful group of sculpture.... – *JHB to Hutchinson, December 4, 1919*

In one report to Hutchinson, Breasted painted a humorous picture of the competition for fine objects:

> These pieces were bought by Dr. Gordon, director of the Philadelphia Museum, but he is not an orientalist and he has now written Blanchard with such uncertainly about them, that Blanchard regards himself as released for Gordon paid no money. An hour ago, I learned of this and mounting a borrowed bicycle for lack of any other conveyance (for I live in a suburb), and the trains are on strike, I rode as fast as I could to Blanchard's place. I saved the bronze by only a few minutes, for Colonel Samuels, a wealthy British officer, was just about to pay the money for the jackal (fig. 4.16). As for the superbly colored relief (fig. 4.17), it will be snapped up the minute the Metropolitan Museum people see it, and they are expected hourly, for they have landed at Alexandria. Under the circumstances, there was nothing to do but buy these two pieces outright, and I have done so, in order to save them for the Art Institute. – *JHB to Hutchinson, December 17, 1919*

The Board of the Art Institute had such faith in Breasted's taste and abilities that in late December they cabled him an additional $10,000 for purchases.

Breasted took the responsibility of adding to museum collections, whether his own or others, very seriously, and he worried about using the best judgment. In one letter, he wrote:

> Then we spent the afternoon with the dealers, who will be the death of me. It's fine to be able to buy after all these years, but o my, the work and the responsibility! Is this bronze falcon at Tano's for 15 pounds as good a purchase as the other one for which Kyticas is asking 20? Would the Art Institute people value a silver bronze statuette of Imhotep more than a fine artist's model of a lion in limestone relief? – *JHB to Frances, January 16, 1920*

---

[6] For the relationship between the University of Chicago and the Art Institute, see Teeter, "Egypt in Chicago."

FIGURE 4.16 Bronze jackal purchased by Breasted from Ralph Blanchard for the Art Institute of Chicago. Dynasty 26, 664–525 BC. AIC Museum Purchase Fund 1920.252. Photo courtesy of the Art Institute of Chicago

FIGURE 4.17 Wall fragment from the tomb of Amenemhet, showing the deceased and his wife Hemet. This relief was purchased by Breasted from Ralph Blanchard for the Art Institute of Chicago. Dynasty 12, ca. 1991–1784 BC. AIC Museum Purchase Fund, 1920.262. Photo courtesy of the Art Institute of Chicago

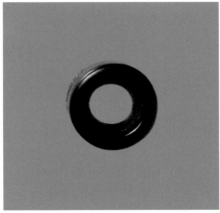

FIGURE 4.18 Seal cylinder purchased by Breasted from Maurice Nahman in Cairo. Incised with the titulary of King Snefru (ca. 2613–2589 BC), but the style of the carving and the way that it is pierced suggest that it is a modern forgery (OIM 10480)

Breasted was acquiring a lot of material in a short time and it was inevitable that a few dubious items were purchased. A perfectly preserved seal cylinder inscribed for King Snefru (fig. 4.18) was purchased from Nahman. Breasted recorded the purchase: "Just as I was leaving [Cairo] I picked up the official seal cylinder of King Snefru, the builder of the first great pyramid ..." (*JHB to Frances, June 15, 1920*). The unusual orientation of the signs and the straight, wide hole bored through its length suggest that it is a modern forgery.

Overall, Breasted spent over $53,000, the equivalent of $675,000 in today's dollars, and many of the objects he bought are key pieces in the Chicago collections.

Although the visits to the dealers dominated his time in Egypt, Breasted intended to scout sites for future excavation. However, his travels in Egypt appeared to be motivated more by social and political ends than scientific. He visited the Giza plateau and nearby Abu Roash with the Allenby party (fig. 4.19), and traveled to Beni Hasan, Tell el-Amarna, and Abydos in Middle Egypt with Mr. Greg, the Director General of the British Foreign Office, and his family. He reported to his wife that they traveled as a group because "It is not safe to go to Amarna alone, and as this is a government party, I thought I ought to go.... There were I suppose a dozen rifles accompanying us" (*JHB to Frances, December 19 and 22, 1919*).

But there were a few outings with his Egyptology colleagues, such as a visit to Saqqara (fig. 4.20) and another to see the excavation of the palace of Merneptah under the direction of Clarence Fisher of the University of Pennsylvania. Breasted commented on the signs of unrest that he saw just outside Cairo and how little the locals had achieved by their resistance to the British:

I took the train for Bedrashein to visit Fisher and his Philadelphia excavations. As a matter of fact it is now impossible to get off at Bedrashein, for the natives of the town formed a mob during last spring's disturbances and burned the station. Since then trains have ceased stopping there, the people of the town have lost all the visitors

FIGURE 4.19 Touring archaeological sites with the Allenby party. The pyramids at Giza can be seen in the distance. December 1919 (OIM photograph P. 25977)

FIGURE 4.20 Touring Saqqara, looking northeast from the Serapeum (the catacomb of the sacred bulls) toward Cairo. Mr. Sanborn, an Egyptologist from Harvard who worked at nearby Memphis, is in the foreground leading a donkey. February 9, 1920 (OIM photograph P. 6931)

> they used to have and thus sacrificed all this business, while to make matters worse for them they and the people of the surrounding villages who took part in the mischief have been heavily fined to rebuild the station and pay other damages. – *JHB to Frances, November 30, 1919*

Other outings included a visit to Saqqara with Egyptologist Cecil Firth and the then recently arrived Ludlow Bull. They spent "a wonderful day among the tombs and pyramids" (*JHB to Frances, January 14, 1920*). The next day, Breasted was joined by British Egyptologist James Quibell to see the recent German excavations of the pyramids at Abu Sir and nearby areas. At Abu Sir, Breasted mused that he was viewing firsthand the source for the pyramid model that stood outside his office door in Chicago (fig. 4.21).

Ironically, the temple of Medinet Habu in western Thebes that was, and continues to be, the concession of the Oriental Institute is not mentioned in his correspondence. However, one can see the inspiration for Chicago House, the headquarters of the University of Chicago in Egypt, and for other lavish accommodations for Chicago excavation teams in Palestine and Iraq from his visit to the French Institute in Cairo. The French Institute was (and still is) housed in a palace built for Princess Munireh, a daughter of the Egyptian khedive or sultan. Breasted described it to his wife:

> It is a huge and sumptuous building, containing large apartments for the Director and the Secretary and Librarian, besides a suite of library rooms, one of which is a spacious hall; also with living and study rooms for six students. This place, with an elaborate printing office for Oriental Languages alongside, is the home of French Egyptology in the land of the Pharaohs. The men in charge and the students have nothing to do but carry on research work. They have enjoyed these facilities for many years; but the substantial returns to science are amazingly meager. I cannot but give my imagination free rein as I dream of what might be done with such an institution with a little vision

FIGURE 4.21 The pyramid of Sahure (ca. 2475 BC) at Abu Sir, taken from the ruins of the Valley Temple, along the causeway. February 9, 1920 (OIM photograph P. 7804)

FIGURE 4.22 Chicago House, the original field headquarters of the University of Chicago's Epigraphic Survey in western Thebes, 1924. The relative luxury of the house was probably inspired by Breasted's visit to the French Institute in Cairo (OIM photograph P. 11143)

and practical ability at the head of it. Why should not our country have a place like this here? If I should spend the next few years devoting all my time and energy to this end, I suppose it could be done. – *JHB to Frances, December 14, 1919*

The description closely matches Chicago House which, in its current form was established in 1930 to replace "Old" Chicago House of 1924 on the west bank (fig. 4.22). It has one of the largest libraries in Upper Egypt, suites and offices for scholars, a formal dining hall, and a large staff of Egyptians to care for the "scientific" staff.

Another aspect of Breasted's research focused on emerging technology — the use of aerial photography in archaeology. The planes of the RAF played an important part in British control over Egypt, and even more so over their mandate in Mesopotamia. Breasted met with Major General Bols, Allenby's Chief of Staff, to request prints of all their air negatives of the Nile Valley, Palestine, and Syria, which one might assume included sensitive or classified images. Later that day, he received a message: "Lord Allenby desires that he be given every facility" (*JHB to family, November 11, 1919*) which he later presented at the Air Headquarters in Zamalek that was quartered in the Gezireh Palace Hotel (now the Marriott). He was assured that he was "to have anything I want among their air negatives of Egypt, Palestine or Syria" (*JHB to family, November 11, 1919*). However, once presented with hundreds of prints, Breasted found that there was "practically nothing of any use to us archaeologists" (*JHB to Frances, December 14, 1919*).

Breasted requested permission take his own photos from an RAF plane. Again, his political connections paid off, for initially the Air Commodore said that civilians were charged 20 pounds ($100, now about $1,275) an hour. Again, Allenby intervened, declaring that the purpose was "scientific work" and it must be done without charge (fig. 4.23). So in January, on a cold, cloudy day, hardly the best for photography, Breasted went aloft. In a letter to Frances, Breasted recalls in great detail the flight and the difficulties in photographing from an open-cockpit plane with a bellows camera (fig. 4.24):

> My pilot climbed down from his machine and brought me a helmet fur-lined, fur-mounted goggles, an air pilot's huge leather overcoat and a large pair of heavy gauntlet gloves…. The young officers crowded around and fastened me into my gear, till I looked like Peary in the Arctic regions…. The pilot disappeared over the top of his covered perch, telling me as he did so that he had fastened a notebook and pencil over my seat and I could write to him all I wanted him to do…. He put on the power and we marched slowly down the field, rolling on the wheels to the other end of the airdrome, so as to turn around and rise facing the wind…. Then with a tremendous roar the machine rushed back across the field again, as the young fellow put on full power, and presently we lifted and were off over the roofs of the hangars and the buildings of New Heliopolis. It was terrific. As we sat directly behind the propeller we received in our faces the full power of the terrible vortex caused by the revolving screw. It was impossible to speak a word in the crashing noise of the engine and the rush of the wind. I opened my mouth to find myself gasping and choking, and quickly perceived that one could only breathe through the nose. But I was seriously asking the question whether I could stand two hours of it, for I saw that the pilot has a glass wind shield and the observer was not protected in anyway, except that he sat deep in his perch.

> We rose rapidly and headed directly westward across the southern apex of the Delta. Then the full splendor of it all broke upon me, and it was thrilling beyond all words to express. Five thousand feet below spread the green carpet of the Delta with the misty wilderness of the desert stretching for a hundred miles on east and west…. Before I knew it we were sailing over the margin of the desert at the western edge of the Delta, and I was looking obliquely down on the ruined pyramid of Aburoash…. I had the camera all ready for the first shot, and when I lifted it above the edge of the car the blast flattened the bellows and drove them into the field of the picture. Do what I would I could not prevent it, and I had to make the exposure anyhow, with much of the view cut off by the intruding bellows. Then the five miles from Aburoash to Gizeh were passed in less than as many minutes and we hovered over the Great Pyramid. I suppose I am the first archaeologist who has ever opened a camera on the pyramid from a point where all four sides could be seen at once….

> The air was very lumpy and at frequent intervals we dropped with a sickening fall into a hole in the air, as you come down in an elevator. This had been going on for nearly an hour. I stuck to my

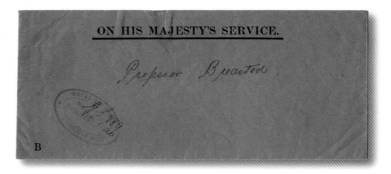

FIGURE 4.23A–B (*A*) Order allowing Breasted to take aerial photos, dated January 10, 1920, delivered in (*B*) an envelope imprinted "On His Majesty's Service"

A

B

FIGURE 4.24 Breasted in full aeronautical dress for this flight over the Giza plateau in a Royal Air Force biplane. January 13, 1920 (OIM photograph P. 68544)

FIGURE 4.25 Aerial view taken by Breasted of the pyramids at Abu Sir. Taking photos with a bellows camera in an open-cockpit plane proved to be very problematic (OIM photograph P. 7797)

pictures and to studying the terrain from one great pyramid cemetery to the next, grinding my teeth and swearing I was not going to give up to it. But it was all of no avail. I leaned over the cockpit rail and surrendered to the Sahara a very good thirty piaster lunch!

... I was not a bit sorry when we turned about and sailed away northward on the return.... The magnificent panorama of the eastern desert illuminated by the low afternoon sun behind us as we swung northward I shall never forget. – *JHB to Frances, January 15, 1920*

Indeed, the challenges of photography under those circumstances resulted in few usable images, for many of them were obscured by the struts of the plane or the shadow of the camera's bellows (fig. 4.25).

BY SEA TO MESOPOTAMIA

Since the land route from Egypt to Mesopotamia was considered too unsafe, Breasted and his team traveled by sea toward Basra in British-controlled Mesopotamia, a journey of nearly two weeks. The trip took them through the Suez Canal (fig. 4.26) and eventually to Bombay, a detour that added nearly 1,500 miles to the most direct route, but one that was necessary because of military permissions — the British army in Mesopotamia was commanded by the British India Office and was indeed composed largely of Indian soldiers, a revealing comment on the way that empires maintain themselves.

Their passage down the Red Sea illustrated Breasted's sense that he was traveling through a landscape that was meaningful because of its past rather than its present:

I have never been through the Canal before, as you know and I naturally found it interesting to pass or rather cross the line of march of Pharaoh's armies in Egypt's great campaigns in Asia. – *JHB to Frances, February 19, 1919*

FIGURE 4.26 The University of Chicago Expedition on board the ship *City of Benares* in the Suez Canal bound for Bombay. Left to right: Breasted, Luckenbill, Shelton, Edgerton, and Bull. February 1920 (OIM photograph P. 6953)

At different points in this journey by sea, his letters also evoke names from antiquity including Moses, Sindbad the Sailor, the mysterious land of Punt, Darius, and Alexander the Great.

Other passengers on this voyage included American businessmen who spent their time drinking and playing cards, American missionaries, and British officers, one of whom allowed Breasted to read a secret report on the activities of T. E. Lawrence in the Arab campaign to take Damascus.

The team spent two uncomfortable days in Bombay, scrambling to find accommodations and berths on a ship to Basra, which were difficult to find because of the increasingly unstable security situation in Mesopotamia. The day before he left, he met with Sir George Lloyd (the British Governor of Bombay),[7] who told him about the difficulties of administering India, and suggesting that the United States should help in this effort.

MESOPOTAMIA

Arriving in Basra (fig. 4.27), Breasted made contact with British military officials who had previous instructions to provide the team with accommodations, transportation, and contacts. They found that the signs of British military occupation were everywhere:

> It was a rapid drive of half an hour, carrying us through various military quarters bearing the names of Arab villages now completely effaced, but once distributed far and wide over the palm-grown plain all around Basra. — *JHB to family, March 14, 1920*

FIGURE 4.27 Ashar Creek in Basra lined with traditional houses of the Ottoman period. March 1920 (OIM photograph P. 7289)

---

[7] Not to be confused with David Lloyd George, who at that time was Prime Minister of the United Kingdom.

To make their preparations to begin camping in uncertain circumstances, they were supplied with a car and a driver — "a turbaned East Indian who knows almost no English" (*JHB to his family, March 14, 1920*).

Breasted was prepared to take the new rail line from Basra to Baghdad with stops to visit the major Babylonian sites. The rail line passed right through Nasiriyah, close to the site of Ur (figs. 4.28–30), known in the Bible as the birthplace of Abraham. The sign at the station marked it as "Ur Junction" (with a helpful Arabic version that did not translate but rather spelled out the English words). On their way to Baghdad they also visited Eridu, Tello, Umma, Uruk, Nippur, Babylon, and Borsippa (figs. 4.31–33). Breasted's letters home convey few of his historical observations of these sites. Those more detailed comments would have been written in his journal, now apparently lost.

FIGURE 4.28 Ur Junction, which Breasted described as "… a group of tents, a mess house, and a row of quarters for the officers of the army in charge, a post office, and three tents in a row serving as a railroad restaurant…." The group arrived by train, then transferred to two cars for the trip to the archaeological site. March 1920 (OIM photograph P. 65833)

FIGURE 4.29 The expedition with members of the British army at Ur with their car and Indian driver. Breasted is in the center looking at the camera. Edgerton and Shelton are to the right. Photo by Luckenbill (OIM photograph P. 7026)

FIGURE 4.30 Breasted at Ur. He commented that it was "the first ancient Babylonian city I had ever visited…." He remarked on the biblical references to the city for the bricks were "marked with the name of Nabonidus, the father of Daniel's Belshazzar…" (OIM photograph P. 65835)

FIGURE 4.31 View of Nippur showing the court of the temple, looking northwest with the great ziggurat (temple tower) in the background with the then abandoned University of Pennsylvania's excavation house on its top. March 1920 (OIM photograph P. 6512)

FIGURE 4.32 The gate at Babylon, showing the towers on the west side. The walls are made of brick with molded images of animals that symbolize the main deities of Babylon. The site transported Breasted to the past as he mused about "… the pavement still in position just as the Hebrew captives must have walked on it." March 1920 (OIM photograph P. 6537)

FIGURE 4.33 View of a sailboat on the Euphrates River at Babylon, taken from the British officers' rest house. April 1, 1920 (OIM photograph P. 7144)

The British military presence was everywhere — the team would encounter biplanes of the Royal Air Force, gunboats, and Rolls-Royce armored cars in addition to soldiers and military camps throughout Mesopotamia (figs. 4.34–36). They had begun a policy of bombing villages in order to pacify tribal groups that were beginning what would turn into a large-scale revolt later in 1920. A British officer named Daly explained to Breasted that

> He could go out 50 miles with his pilot and bomb a tribe, come back for the usual morning's work at his desk; run out and give 'em another after lunch and transact the regular afternoon's business before tea, or postpone the bombing picnic until <u>after</u> tea, and return in plenty of time for a bath before dinner. The scattering on camels the first time he did this, said Daly, was very amusing. In two cases the tribal sheikh held out for fifteen days and then yielded to the discontent of his tribe and came in and submitted. – *JHB to Frances, March 30, 1920*

FIGURE 4.34 Breasted in the rear passenger seat of a British R.E. 8 biplane at Abu Kemal. He and Luckenbill had the opportunity to go aloft to follow the Euphrates and view the desert formations from 2,000 feet. May 1920 (OIM photograph P. 7346)

FIGURE 4.35 A British gunboat on the Euphrates River at Falluja (OIM photograph P. 7297)

FIGURE 4.36 Fleet of Rolls-Royce armored cars fitted with fortified gun turrets in the British camp at Abu Kemal (OIM photograph P. 7345)

This policy would affect Breasted and his team directly in what could have been a fatal encounter, the first time Breasted mentions directly interacting with local people in Mesopotamia. The group was visiting Umma (modern Tell Yokha), an isolated if very important Sumerian site that they could only reach after a five-hour ride on horseback. They were traveling with sixteen riders including the British Political Officer of the area, Captain Crawford, as well as a number of local tribal sheikhs. As Breasted described it,

> The mounds of Yokha are of vast extent. The strong north wind was driving the sand into our eyes, and besides this, it was too late to make our usual sketch plan of the extent of the ruins. Shelton had managed to hang pretty close to our flanks, and he soon came over to us. A moment later he plucked my sleeve and said, "Who are all these?" Looking where he pointed, I saw a body of 30 or 40 Arab horsemen, sweeping up the slope of the mound directly upon us. Crawford was 50 paces away and did not see them. I walked over to him and asked him to look round. His face never changed, and with the utmost composure he asked our Arabs who these horsemen were. They replied they were the Bnê Ghweinîn.[8] In a moment they halted, drawn up in an impressive line, like a platoon of cavalry on parade. The Bnê Ghweinîn had been recently bombed by British airmen; their sheikh and many of his followers had been outlawed, and these were the men before us, a hundred paces away.... Crawford said afterward, "I thought we were surely done in...." [fig. 4.37]

> ... Four sheikhs dismounted from their horses, left them in the line and came forward to us. The sheikhs in our party introduced them and they all stepped forward and kissed Crawford's right shoulder, at the same time dropping from their heads their rope-like *agâlas* arranged in coils over their headcloths. To let the *agâla* fall thus to the shoulders is a token of complete submission. It was quite evident that this had all been arranged beforehand by the sheikhs who accompanied

FIGURE 4.37 Members of the Bnê Ghweinîn tribe under "the outlaw sheikh" Mizal "with his fellow out-laws" posing for a photo at Umma with British Political Officer Captain Crawford (OIM photograph P. 6750)

---

[8] It has not been possible to verify this name or relate it to a modern tribal group in the area.

FIGURE 4.38 Feast of Sheikh Mutlaq, the successor of Mizal, at Kalat es-Sikkar (OIM photograph P. 6753)

us. Crawford told Sheik Miz'al he must come along with him to Kalat es-Sikkar and afterward to headquarters at Nasiriya to make his formal submission there and stand his trial for his misdeeds. Miz'al was not expecting this and the palaver which followed was long and interesting as one sheik after another took up the word. Miz'al did not assent but rode with us nevertheless to the tents of his tribe, — a two hour's ride eastward toward the river, the Arabs shouting, racing at wild speed, caracoling their horses in wide curves and brandishing their rifles.

Here we arrived about 4:30 p.m. and were at once taken to the *madhif* or guest-tent of Sheik Mut-laq, Miz'al's brother who is now sheikh in Miz'al's stead. The big black camel's hair tent, open on one side, was carpeted with gay rugs and at the right were cushions where Crawford and I seated ourselves, the rest of our party on our right, then the sheikhs who were with us, and the notable men of the tribe [fig. 4.38]. Tea and cigarettes were at once brought in and passed by Sheik Mut-laq himself. Then four men appeared carrying between them an enormous tray heaped high with boiled rice on which lay two whole roast sheep. It was set down in the midst and a smaller tray of rice, together with numerous roast chickens, pieces of roast mutton, bowls of clabbered milk and generous piles of Arab bread, were all placed before our party. As we fell to, the leading sheikhs gathered round the big tray, and a circle of dark hands carried the food to a circle of dark faces in a scene which I am really too tired to describe. The food was really well cooked and delicious.... — *JHB to Charles, March 24, 1920*

On their way to Baghdad, the team visited the Shia holy cities of Karbala and Najaf. Breasted noted an unusual local practice (fig. 4.39):

I have met a man carrying a corpse wrapped in reeds and balanced across his horse on the pommel of his saddle, while he rode behind it and kept it in equilibrium as he followed the winding road across canals and embankments. Such "corpse-carrying" is widely practiced and there are men who follow it as a calling.... These bodies are brought from far and near for burials at Nejef by the tomb of Ali. – *JHB to Frances, April 4, 1920*

FIGURE 4.39 Corpse carrier taking a body for interment in Najaf. The body, wrapped in reeds, lies across the pommel of the horseman's saddle (OIM photograph P. 6799)

FIGURE 4.40 The entrance of the tiled mosque of Imam Ali in Najaf, with merchants selling goods (OIM photograph P. 7164)

The mosque of Imam Ali in Najaf was particularly beautifully tiled (fig. 4.40). Breasted was inclined to see it, and in particular the vendors in front of the mosque, in terms of a direct connection to the ancient past:

> all sorts of merchandising is carried on — vendors of fish and vegetables, among them many women, jostle the low stands of the squatting money-changers, sitting in rows along the walls of the courts, and one is forcibly reminded of Jesus' cleansing of the temple in Jerusalem. — *JHB to Frances, April 4, 1920*

Arriving in Baghdad (figs. 4.41–43), Breasted met with the highest-ranking British officers — General Haldane, the Commander in Chief of British forces in Mesopotamia, as well as (more importantly) Major General Percy Hambro, the Quartermaster General who controlled all transportation in the country, whom Lord Carnarvon had previously contacted on Breasted's behalf. He also met with the American Consul, who was responsible for

FIGURE 4.41 The Tigris River at Baghdad. Note on the right a bridge made of small boats (OIM photograph P. 7254)

FIGURE 4.42 A greengrocer's shop in Baghdad (OIM photograph P. 7257)

FIGURE 4.44 Aqar Quf, the site of the ruined ziggurat of Dur Kurigalzu (fourteenth century BC), which in earlier times had been mistakenly identified as the ruins of the Tower of Babel. Right to left are Gertrude Bell who would later help to found the Iraq National Museum, Major General Percy Hambro, Daniel Luckenbill, and Sir Hugh Bell, Ms. Bell's father (OIM photograph P. 6814)

FIGURE 4.43 A view of the *suq* or bazaar in Baghdad (OIM photograph P. 7258)

FIGURE 4.45 The ruins of the east façade of the magnificent Sassanian palace of the third century AD at Ctesiphon, the largest standing mudbrick arch in the world. Breasted and his team visited the site in the company of the British Chief of Staff. Their cars can be seen in front of the palace. April 1920 (OIM photograph P. 6556)

mail delivery and who also provided introductions to local antiquities dealers from whom Breasted was to purchase cuneiform tablets and a clay prism with the inscribed Annals of the Assyrian king Sennacherib. Breasted was concerned about whether he would get official permission to bring these antiquities out of British-controlled Mesopotamia:

> The Civil Commissioner who is practically king of the country, Colonel A. T. Wilson, is an exceptionally strong man; ~ has very decided views about the proper policy for his treatment of the country, and I fear has such a sensitive regard for what he considers the rights of the Arabs (imagine Arab rights based on their appreciation of cuneiform documents!) that he may not allow us to take out a single thing. – *JHB to Frances, April 9, 1920*

Among the British officials whom Breasted met was Gertrude Bell, a woman who had lived and worked in the Middle East for about five years and was (in Breasted's words) "a terrible blue stocking" (*JHB to Frances, April 10, 1920*), that is, an educated woman who may not have been concerned about her appearance. They visited the Babylonian ziggurat (temple tower) at Aqar Quf (fig. 4.44) on a day trip from Baghdad along with Bell's father and Major General Hambro, and (according to *Pioneer to the Past*), had a late meeting with a tribal sheikh to secure his loyalty. Breasted also visited the magnificent Sassanian palace at Ctesiphon (fig. 4.45).

Throughout his journey, Breasted again crossed paths with the Assyriologist Albert T. Clay of Yale, who was following virtually the same itinerary and schedule, through Europe and Egypt to Mesopotamia and ultimately to Jerusalem. Clay was primarily looking for cuneiform documents to purchase and Breasted clearly struggled to get along with him (and indeed would later defend himself against Clay's accusations that Breasted bought tablets in Baghdad that had been promised to Yale (*correspondence between JHB and A. T. Clay, 1921*).

From Baghdad, the expedition traveled by the newly inaugurated Mesopotamian Railways (fig. 4.46) to visit the Assyrian capital city of Ashur, then by armed caravan through dangerous territory to the area of Mosul, where they could visit other Assyrian sites: Nineveh (fig. 4.47), Nimrud, Khorsabad, and the temple at Balawat. This portion of the trip was dangerous, the rail line being cut while they were in the north, and Breasted heard accounts

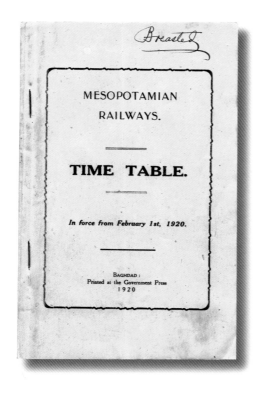

FIGURE 4.46 Schedule for the Mesopotamian Railways, 1920

FIGURE 4.47 View of the lower town of Nineveh, today within the suburbs of the modern town of Mosul. The photo was taken from the citadel mound called Kuyunjik, along the ruins of the ancient city wall. The mound in the background is the site of a shrine to Nebi Yunus (the prophet Jonah) as well as an earlier Assyrian palace. A tent encampment of the British army is visible at left (OIM photograph P. 6581)

FIGURE 4.48 Ruins of the walls and gateways of the northwest quarter of Nimrud. Note the crowned head of the massive, human-headed winged-bull guardian figure emerging from the debris. April 1920 (OIM photograph P. 6568)

FIGURE 4.49 An unidentified member of the expedition standing on the reliefs of the Assyrian king Ashurnasirpal II (883–859 BC) emerging from the ruins of Nimrud. The ziggurat rises in the background. April 1920 (OIM photograph P. 6571)

FIGURE 4.50 *Stonecutters in the suq in Mosul. April 1920 (OIM photograph P. 6841)*

of British officers who had been ambushed and killed in the area. Yet these sites clearly excited Breasted's greatest interest, with massive carved-stone reliefs and winged bulls visible protruding from the earth (figs. 4.48–49), and he made preliminary plans to request permission to excavate at Nimrud from British officials (in part to forestall the proposals of Clay from Yale University).

Wandering in Mosul, Breasted once again construed the present as unchanged for thousands of years when he saw stonecutters working in the bazaar (fig. 4.50):

> Here are the same crafts and the same tools which enabled the Assyrian Emperors to build their palaces across the river, 2700 to 2600 years ago. I found workmen sawing up blocks of alabaster, just as their ancestors did to furnish the slabs for the splendid sculptured relief wainscoting which lined the magnificent halls of the Ninevite palaces, when Sennacherib was besieging Jerusalem, and Isaiah was delivering political speeches on the street corners twenty-six centuries ago. – *JHB to Frances, April 19, 1920*

Upon returning to Baghdad, Breasted met Colonel A. T. Wilson, the Civil Commissioner (the highest civilian official) in Mesopotamia. Breasted wrote:

> I promised the Civil Commissioner at Baghdad to hand him a complete plan for the organization of a Mesopotamian Department of Antiquities. What is more, if I could put my hand on young Americans of the right experience, I could also man the organization for him, and he would be very glad to get them, for there are no English Assyriologists. – *JHB to Frances, June 19, 1920*

Wilson asked Breasted to alter his travel plans in order to investigate some wall paintings that had been discovered at Salihiyah by troops digging a machine-gun emplacement up the Euphrates, in the midst of an area in open revolt. Breasted altered his plans of returning to Egypt by sea (and of visiting Persia, specifically the trilingual cliff inscription of Darius at Behistun) and decided to attempt to cross the Syrian desert.

FIGURE 4.51 The University of Chicago Expedition caravan of Fords traverses the desert near Falluja, bound for Abu Kemal (OIM photograph P. 7299)

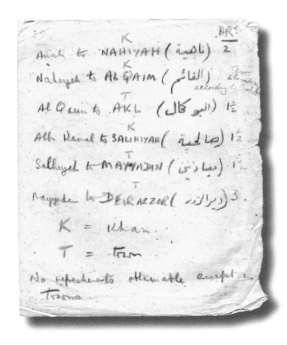

FIGURE 4.52 Copy of handwritten directions from Anah to Deir ez-Zor, with an annotation of travel time in hours, availability of food, and the name of each town written in English and Arabic

The route from Baghdad to Salihiyah would take five days and traversed a particularly dangerous area that was in open revolt. Breasted was concerned about this route as well as the idea of continuing beyond the zone of British control into the Arab State and ultimately to Aleppo.

It is a grave responsibility to take four men beside myself across four hundred miles of war zone, three fourths of which or nearly so are beset by treacherous Arabs. It is likewise to be carefully considered whether a man with a family waiting for him at home ought to undertake such a journey. – *JHB to Frances, April 25, 1920*

The expedition was given a procession of five vans and two touring cars to make the journey (fig. 4.51) and handwritten directions that indicated the major stops along the route (fig. 4.52). This early portion of the journey required a round-the-clock guard, and Breasted and his team took turns: "every fifteen minutes I make the round of the camp with my gun in one hand and Ludlow's *shillelah* in the other" (*JHB to Frances, April 30, 1920*).

Along the way, they suffered from repeated mechanical trouble, as the cars had to cross the rough terrain with no paved road; they had to abandon two cars that broke down. The tension and the difficulty in controlling the caravan finally led Breasted to lose his temper with one of the drivers who was not doing as Breasted wanted:

> I stepped over to him and ... hauled back and slugged the Indian a first class smash in the jaw! At this juncture his remarks were numerous and fairly audible, but ever since he has been quietly on his job. – *JHB to Frances, April 30, 1920*

They stopped at a British camp in Ramadi (fig. 4.53) and visited the natural seeps of bitumen (a kind of asphalt or tar) that had been used in Mesopotamia since antiquity (fig. 4.54):

> I decided to remain at Hît, where we had opportunity to visit the weird bitumen pits or fountains in a basin so wild and desolate that it seemed the very gates of hell with fumes of sulphur, blackened rocks and pits of boiling bitumen. – *JHB to Frances, April 30, 1920*

FIGURE 4.53 British officers who hosted the University of Chicago group in Ramadi. The group at right (left to right) are Breasted, Edgerton, and Shelton (OIM photograph P. 7301)

FIGURE 4.54 Seeps of bitumen (asphalt) at Hit on the Euphrates River, a source of bitumen for Mesopotamians for thousands of years, April 1920 (OIM photograph P. 7309)

FIGURE 4.55 View of the Euphrates at Anah with a waterwheel in the center of the near shore, May 1920 (OIM photograph P. 7332)

FIGURE 4.56 View of the Euphrates at Anah. The waterwheel is visible just off the near shore; the abandoned Ottoman Turkish fort is on the far shore. Photo taken from the British camp (OIM photograph P. 7333)

After four days of difficult and anxious traveling, they arrived in Anah, a village with a British outpost and an abandoned Turkish fortress (figs. 4.55–56):

> You can't imagine what a relief it has been to reach this beautiful, palm shaded village stretching nearly five miles as it struggles along the Euphrates filling entirely the narrow margin between the cliffs of the desert plateau and the shores of the river below.... – *JHB to Frances, May 2, 1920*

They then joined with a military convoy of thirty-four cars to make the final stage of the trip to Abu Kemal, across the most dangerous stretch of track. Breasted noted that convoys in this area were frequent targets of Arab snipers and related an extremely hazardous event that had taken place recently along the road:

> Not long ago they shot the Indian driver of one of the vans right through the heart as he drove his machine at full speed. A neighboring driver sprang out of his machine, stopped the driverless car which was running wild, and saved the body of his comrade, fighting off the Arabs as he did so. – *JHB to Frances, May 4, 1920*

Reaching the large British garrison at Abu Kemal at last (fig. 4.57), they were greeted by Colonel Leachman, the Political Officer in this dangerous area, known to be among the British officers most knowledgeable about Arab culture and language. Although Breasted could not have known it at the time, Leachman was to be killed in an ambush by local tribes three months later, further illustrating the very real dangers in the area.

The British were in the midst of negotiations with the Arab State over this sensitive border area, and in fact were just about to abandon their outposts at Salihiyah and Abu Kemal, which meant that Breasted had just one day to document the wall paintings. These turned out to be paintings from a Roman fortress of the third century AD called Dura Europos. Breasted photographed them and later colorized the images (figs. 4.58–59). Although Breasted asked that the paintings be covered over by refilling the pit, the dirt settled with rain and the paintings were damaged by local people before excavations began just two years later. They are now in the Damascus Museum.

FIGURE 4.57 View of the British army camp at Abu Kemal. The photo, taken from the office window of Colonel Leachman, shows the many rows of horses for the soldiers, May 4, 1920 (OIM photograph P. 7341)

FIGURE 4.58 Photo of a wall painting at Salihiyah (ancient Dura Europos) showing the Roman tribune Julius Terentius offering incense to local gods. Breasted and his team had only a day to document the painting because the British were pulling their troops from the area (OIM photograph P. 6659)

FIGURE 4.59 Colorized photo of the wall painting at Dura Europos as published in Oriental Institute Publications 1, plate 8; note that the turbaned figure in the foreground of figure 4.58 has been painted out, but Breasted (in lower right corner) was left in the photo (OIM photograph P. 6659)

THE ARAB STATE AND BEYOND

After their work at Salihiyah, Breasted and his team were forced to return their cars to the retreating British and exchange them for local wagons for an eight-day journey within the Arab State, following along the Euphrates River and then turning west for Aleppo. Indeed, as they woke up after their first day of traveling, Breasted heard

> a great clatter of horsemen in the street below, and looking out I saw several field guns and a body of cavalry marching down the street. They were going to take possession of the new territory evacuated by the British. – *JHB to Frances, May 6, 1920*

Political relations between officials of the Arab State, local tribes, the French, and the British were extremely tense. At the suggestion of Colonel Leachman, to make sure that there was no mistake about the nationality of the wagon train, Breasted put an American flag on their wagon (figs. 4.60–63).

The image of the wagons flying an American flag evokes pioneers traveling in the American West, and it is noteworthy that there was an American flag ready at hand at the right moment in Salihiyah. Oddly, it is a 37-star flag, which represented the number of American states up until 1877, so the flag was more than forty years old in 1920. As much as the flag conveyed the message "don't shoot," it also certainly established an American presence.

Their journey to Aleppo passed first through Deir ez-Zor (fig. 4.64), where Breasted met with officials and officers of the Arab State, who discussed the local political situation and expressed their dislike of British rule and their hope that America would be able to support them. Breasted, like many travelers, was put in the position of representing his own country, but found it difficult to convey his sense of American politics (not surprising given the cultural and linguistic barriers to understanding). Leaving Deir ez-Zor, the party passed a camp of the Albu Hayyal tribe (fig. 4.65) and were

FIGURE 4.61 Headquarters of the University of Chicago Expedition at Meyadin in the Arab State, with American flag. Left to right: Edgerton, Shelton, Breasted, Bull (in pajamas). Note the pistol on Breasted's belt (OIM photograph P. 7377)

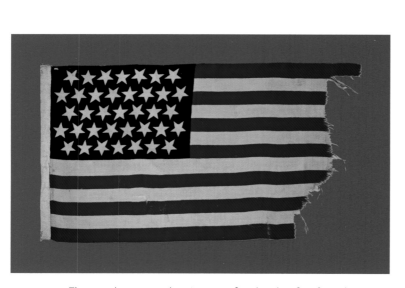

FIGURE 4.60 The expedition carried an American flag that they flew from their wagon when traveling in the Arab State. It is not known whether they brought the flag with them or acquired it in the Middle East (OIM photograph D. 9011)

FIGURE 4.62 The wagon ("*arabanah*") of the University of Chicago at Tibni, Iraq, looking toward the Euphrates. The American flag on the wagon was an attempt to inform the local peoples that the expedition members were American, not British. May 7–8, 1920 (OIM photograph P. 6860)

FIGURE 4.63 A local Arab playing pipes for Breasted (in wagon) at first stop out of Salihiyah (OIM photograph P. 7373)

FIGURE 4.64 View of a camel caravan in Deir ez-Zor, May 1920 (OIM photograph P. 7400)

FIGURE 4.65 The University of Chicago team visiting the Albu Hayyal tribe. Sheikh Ramadhan stands to the left facing his tent. Breasted, Shelton, and Bull stand to the left (OIM photograph P. 7408)

invited to coffee, where Sheikh Ramadhan said that all Arabs love America and asked Breasted to deliver a letter for him to Aleppo.

Although Breasted didn't know it at the time, by agreeing to carry the letter, he had been drawn into a complex and potentially dangerous political negotiation, about which he only learned months later:

> Ramadhan was an officer in Lawrence's army, and lost part of his nose there.... After the armistice King Faisal gave him large sums of money to carry on propaganda in the north among the Kurds and Turks on his behalf. It was then discovered that he was carrying on propaganda in the interests of Mustapha Kamal, the renegade Turk (really a Saloniki Jew),[9] who is head of the rebellious Young Turk party in Asia Minor and is leading a powerful army there. Faisal then recalled Ramadhan and afterward made him governor of Der el-Zor, where he led the Arab seizure of the place.... Now he is again flirting with the Turks of Mustapha Kemal's outfit, but this time seemingly in the interests of the French! ... He is working against Feisal's interest again, and made me his messenger to the French without my knowing it! – *JHB to Frances, June 10, 1920*

Although their route took them past many important archaeological sites, the area was little known to archaeology at the time, and the party apparently did not visit any ancient mounds. Breasted did note an important mound, inaccessible on the other bank of the river:

> The river plain all along this region from Sabkha up to El-Hammam is wider than in the Hît to Anah region. A small nation is thinkable in this region, and indeed on the opposite (left) bank of the river, at the mouth of its northern tributary, the Balikh, there are extensive ruins of an ancient town, which has never been more than cursorily examined. At present the place is called Rakka, and fine Neo-Persian blue-glazed bowls are found there and sold by the European antiquity dealers. We were sorry that we could not cross and examine these ruins, but the other side of the river is very unsafe. – *JHB to Frances, May 9, 1920*

Rakka (Raqqa) was an important center during Early Islamic times; the mound was certainly Tell Bi'a (ancient Tuttul), an important political capital and cult center for the god Dagan, although this was not known at the time.

The following week was mostly notable for the extreme discomfort of the wagons themselves and of the caravanserais or khans (fig. 4.66) in which they stayed each night.

> The rooms in these khans are raised a story above the barnyard enclosure where the horses, wagons and drivers spend the night. But in this hole there is only one second floor room and that was taken. We are therefore down with the horses and all the rest of the mess just outside our door. This might be thought to possess something of a rustic flavor not altogether intolerable, but for the fact that a heavy wind is blowing. It picks up all the dried horse droppings outside and they become droppings in a very personal sense.... The Babel of arriving and departing wagons ... continued without interruption all night: — drivers shouting to their comrades and their horses, wagons rattling: — in short a constant row which beggars description.... We are simply alive with fleas.... Tonight in this execrable den of filth we are 25 miles from civilization and a comfortable hotel in Aleppo. – *JHB to Frances, May 11, 1920*

Breasted was greatly relieved and proud to have finished that portion of his journey — as he notes at several points, he believed his group to have been "the first white men to cross the Arab State since it was proclaimed" (*JHB to Frances, May 13, 1920*), although another American archaeologist, Francis Kelsey of the University of Michigan, had traveled between Aleppo and Damascus earlier that same year (Thomas, *Dangerous Archaeology*).

---

[9] Also known as Atatürk, he would found the Republic of Turkey in 1923; Breasted was misinformed about Mustapha Kemal's religious background.

FIGURE 4.66 View of the caravanserai in es-Sabkha where the group and their wagons spent a night. As Breasted described it, "a pig-sty would be palatial beside the filthy hole in which we are preparing to spend the night! Nothing I have ever seen approaches it," May 1920 (OIM photograph P. 6891)

FIGURE 4.67 Group of Armenians gathered at the United States Consulate in Aleppo, waiting for visas or passports (OIM photograph P. 7468). The photo is poignant since hundreds of thousands of Armenians had died in the preceding five years, many had fled Armenia, and the United States was considering making Armenia a protectorate or mandate as Breasted took this photo

FIGURE 4.68 View of the heavily fortified entrance to the Citadel at Aleppo, May 1920 (OIM photograph P. 7470)

FIGURE 4.69 Breasted and the governor of Aleppo in the entrance to the Citadel. He gave Breasted helpful letters of introduction for his visits to Kadesh and Baalbek. Breasted commented that the governor was, unlike most locals, interested in the ancient history of the city. May 1920 (OIM photograph P. 7471)

Arriving in Aleppo, the party stayed at the Baron Hotel (still open for business today). Breasted had a high fever, although he did not mention this in his letters home. Years later and after repeated bouts of illness, it was found that he had contracted amoebic dysentery (C. Breasted, *Pioneer to the Past*, p. 300).

He met with the American consul and the Arab governor (figs. 4.67–69) and discussed the unstable security situation. Breasted ultimately decided that he did not feel safe visiting two important sites in the area: the important Hittite city of Carchemish (then as today right on the Turkish border), and the Roman city of Antioch in the Amuq Plain on the Mediterranean coast (where the Oriental Institute would work in the 1930s). With an armed escort provided by the Governor of Aleppo, they were able to visit the site of Kadesh (fig. 4.70), south of Aleppo and the location of a battle between Egyptian King Ramesses II and the Hittites in the thirteenth century BC that Breasted had studied, and then visited the Roman city at Baalbek, in what is now Lebanon.

From Baalbek, the expedition visited throughout what is now Lebanon, as security concerns allowed, from a base in Beirut. Harold Nelson, one of Breasted's

FIGURE 4.70 View of the mound at Kadesh, looking west toward the Orontes River. In 1903, Breasted had published a monograph on the ancient battle of Kadesh and so the site was of special interest to him. Other members of the University expedition are in the foreground, May 1920 (OIM photograph P. 6866)

FIGURE 4.71 View of the American University of Beirut, the former Syrian Protestant College, from West Hall, showing the rooftops of the campus looking out on the Mediterranean to a steamship in the bay. Breasted commented on the importance of the University: "The graduates among the native Syrians are scattered all over the Near East. Two of them are in Faisal's cabinet at Damascus, and the college has to be careful lest it should be compromised by the political activities of its former students who are now of course ardent nationalists loudly demanding independence" (OIM photograph P. 7545)

FIGURE 4.72 View of the American University of Beirut, with its beautiful campus on the Mediterranean. College Hall is in the center, Assembly Hall to the left (OIM photograph P. 7609)

FIGURE 4.73 The expedition in cars traveling along the Mediterranean coast north of Tripoli, May 1920 (OIM photograph P. 6874)

FIGURE 4.75 The walls of the Nahr el-Kelb Valley are incised with historical inscriptions from armies who traversed the area over thousands of years. The large rock-cut stela protected by shutters commemorates an ancient Assyrian victory. Above it is a commemoration that the "British Desert Mounted Corps aided by the Arab Forces of King Hussein captured Damascus, Homs and Aleppo, October 1918." May 1920 (OIM photograph P. 7508)

FIGURE 4.74 The Dog River (Nahr el-Kelb) Valley, an important landmark between the mountains and the coastal plain, today in Lebanon (OIM photograph P. 7505)

students and later the first director of the University's Epigraphic Survey in Egypt, was then teaching at the American University of Beirut (figs. 4.71–72) and gave them both a place to stay and information on the political situation. The expedition visited sites along the coast (fig. 4.73), including Byblos and the Phoenician city of Sidon; the Nahr el-Kelb (or Dog River [figs. 4.74–75]), where conquering armies from Assyrians and Egyptians to British and French had left commemorative inscriptions; and a crusader castle just outside the modern town of Tripoli.

The expedition had continued to acquire antiquities along the way, although in limited quantities. While visiting Sidon (fig. 4.76), they stayed with an American missionary, Dr. George A. Ford, who offered Breasted an impressive collection of Phoenician sculpture and sarcophagi for $25,000 "as a contribution to the orphanage of the mission at Sidon." Breasted did not seem to have the energy to purchase the collection then (it was being considered by the American University), and although he pursued it for years afterwards, it ended up in the National Museum in Lebanon.

FIGURE 4.76 View east from the Eshmun temple, "the only surviving portion of a building of any extant in all Phoenicia," at Sidon on the coast of Lebanon. The walls of the platform are to the right. Breasted was very interested in starting excavations at Sidon, but did not end up doing so. May 1920 (OIM photograph P. 6918)

FIGURE 4.77 Once in Damascus, Breasted had two audiences with King Faysal. In this view of Damascus, the Kings' residence can be seen at left center (OIM photograph P. 7576)

Breasted continued making political contacts at this late stage in the journey. He met in Beirut with General Gouraud, the High Commissioner of France in Syria, and then traveled to Damascus to observe a meeting of the Chamber of Deputies of the Arab State and to meet twice with King Faysal — once at his private home, and once for dinner in the palace (fig. 4.77).

> The dinner with King Feisal was very interesting.... We drove to the palace, where we passed through an endless array of sentries, aides de camp, adjutants, chamberlains, the last of whom filled three anterooms in succession.
>
> ... So the Consul and I sat facing each other on the King's right and left. Next to me on my right was the King's brother, Zaid Pacha, and on his right Nuri Pacha. Opposite these two and myself, were the consul and a chamberlain, while at the end of the table facing the King was an officer whom I do not recall as to name and position.
>
> The dinner was simple and about such as you would find in a fair hotel; ~ ending with the famous Damascus pastry and really luscious Damascus fruit. Politics had been rather delicate ground ... the King said bluntly his present unhappy position between French and English aggression, the one in Syria, the other in Palestine, was our (America's) fault! The Consul and I had both demurred, but I do not think it made much difference in the King's feeling.... After dinner the King led the way to a balcony overlooking the palace gardens and the entire city. There was a full moon, and below us lay the gardens of Damascus, the minarets and the sea of houses bathed in bright moonlight! It was a spectacle never to be forgotten.
>
> ... [Because it was Ramadan,] we took an early departure. Before doing so I took from my pocket a photograph of the King, which I had secured for the purpose and asked him if he would be kind enough to sign it. He took it at once to his desk and put on his name in red ink (fig. 4.78). – *JHB to Frances, June 3, 1920*

They did not realize how close they had come to being caught up in warfare; within two months of their departure from Damascus, General Gouraud would order his soldiers to Damascus to remove Faysal from power.

Taking a train to Haifa, Breasted noted evidence of the political turbulence of the moment, as a group of Bedouin soldiers loyal to Faysal took the train cabin Breasted had reserved, forcing Breasted and his team to ride in a third class cabin (which had been emptied of everyone else to make room for them):

> Everybody is in terror of the Beduin, and their services in the war make them a strong group over against the townspeople and the educated modern class. The Beduin terrorize the towns much as did the cowboys of a generation ago on our own frontier in the western states.
>
> We had a long weary ride in the cramped, hard, wooden benches of the third class, with many vermin-infested natives trying to climb in with us at every station.... – *JHB to Frances, June 3, 1920*

Further evidence of turmoil was visible as the train crossed from the Arab State into territory occupied by the British military:

> Hanging from a telegraph pole beside railway line, we saw swaying in the wind the body of one of the Bedouin who had been cutting the Haifa-Damascus line! He had shot two Jews and resisting arrest, he had been properly quieted by the Indians [i.e., British soldiers from India] sent to bring him in. – *JHB to Frances, June 3, 1920*

From Haifa, Breasted hoped to visit Megiddo, particularly interesting to him as the site of a victory by the Egyptian pharaoh Thutmose III over Canaanite forces. The trip was a comedy of errors, with bad directions, mechanical breakdowns, and careless driving preventing them from getting more than a distant glance at the mound (fig. 4.79). But Breasted was weary after the long voyage and was not as determined as he might have been six months earlier:

> I confess, I am chiefly interested now in getting home. We have accomplished all we have set out to do, except a more full and satisfactory examination of Syria. For this exception the French oc-

FIGURE 4.78 A stock photo of Faysal that Breasted asked him to sign (OIM photograph P. 8247)

FIGURE 4.79 View of the mound of Megiddo from a distance "commanding the pass across the Carmel ridge." The Oriental Institute would excavate at Megiddo from 1925 through 1939. May 1920 (OIM photograph P. 7597)

cupation is responsible. As far as museum acquisitions are concerned, I have every reason to feel contented. We shall be able to make a creditable showing, and one that will not fail to bring in more funds for Oriental work. – *JHB to Frances, June 3, 1920*

In Haifa, the team began to break up. Luckenbill and Harold Nelson returned to Beirut, where Luckenbill was to develop the photographs taken on the journey. This marks the end of the photographic record from the trip, and Luckenbill would return to the United States directly from Beirut. Breasted scarcely mentions the others in his letters beyond Haifa, but they returned to the United States from Cairo separately from Breasted.

In Jerusalem, Breasted met scholars at the British School of Archaeology as well as the French Institute, and once again met Professor Clay of Yale. His trip through the Arab State was also of great interest to British officials, and he met both with the Commander in Chief Major General Bols and the Intelligence Chief to describe his impressions of Faysal.

> Bols asked me with much seriousness whether I thought Feisal was really in control of his Arabs. I am confident that in this question lies the chief English difficulty. They have long been subsidizing Feisal and made no secret of doing so, but now that is <u>supposed</u> to have ended. I would be willing to wager a good deal that on the quiet they are still subsidizing him, and what they are anxious about is whether in holding him loyal to the English, they are at the same time holding the Arabs.
> – *JHB to Frances, June 5, 1920*

He was also witness to the unrest caused by British support for the creation of a Jewish state in Palestine, and in the same letter writes:

> The French force themselves into Syria as the unwelcome lords of the Syrians, and produce a situation of growing trouble and disorder; while the British, welcome rulers of Palestine, force upon the protesting people of the land an utterly abhorrent Jewish supremacy, producing in Palestine

a situation equally full of trouble and disorder! It is easy to say that a tottering group of British politicians have been bought up by Jewish money, but I hate very much to believe it. — *JHB to Frances, June 5, 1920*

Unfortunately, the security situation prevented even a visit to nearby Jericho, and Breasted took the train to Cairo five days later.

In Cairo, his final preparations for returning to the United States involved arranging transportation and visas for himself and his team, as well as for a young man named Ali, whom Breasted had hired as cook in Basra, and who had been "as devoted as a faithful dog" through the rest of the journey. Breasted had in mind that Ali would come back to America to help Frances around the house. He was also concerned to arrange insurance and shipment of the antiquities he had purchased six months before, but instead got a request from Lord Allenby:

> I walked into the Residency to secure some help in a matter of baggage at the Custom House, and I came out charged with an international mission which may have something to do with saving Palestine from civil war, and the whole Near East from a conflagration. — *JHB to Frances, June 10, 1920*

> Lord Allenby took me aside and charged me <u>again</u> to tell the Prime Minister and Earl Curzon all the facts, especially those which would reveal the hostility of the western Arabs to the British, who used to be so popular among them. "I am confident," said he, "that they will listen to you, who are without prejudice, and have no interest to serve, much more readily than they will listen to me." — *JHB to Frances, June 15, 1920*

> "It is of the highest importance that the facts you have told me this evening should be plainly brought before the Prime Minister and Lord Curzon, and you have the opportunity to do a very important piece of work. For they do not realize the situation at all.... They do not understand that the Arabs and the Christians are now united against the Jews and that the present policy is aggravating this anti-Jewish hostility to a dangerous degree. Do not fail to make this clear to them as you have done to me. And above all tell them of the danger of Arab union with Bolshevism in the north, as you told me this evening." — *JHB to Frances, June 16, 1920*

Breasted returned to England and did indeed deliver his testimony to the British government. Although the Prime Minister was away, he met with the Foreign Minister, Lord Curzon, and told him about his firsthand knowledge of the Arab State as well as Palestine. Archaeology had by this point faded completely from the picture — this was pure political intelligence. After a final dash to Paris in pursuit of antiquities — which did not result in a purchase — Breasted returned to the United States, arriving home with Ali in July 1920, eleven months after he had started.

5. THE CHANGING LEGAL LANDSCAPE FOR MIDDLE EASTERN ARCHAEOLOGY IN THE COLONIAL ERA, 1800-1930

MORAG M. KERSEL
JOUKOWSKY INSTITUTE FOR ARCHAEOLOGY AND THE ANCIENT WORLD,
BROWN UNIVERSITY

Ownership of antiquities has for millennia been a potent symbol conveying power and wealth, education, national pride, and ultimately, control of the past. During the nineteenth and early twentieth centuries, the gradual dissolution of the Ottoman Empire and the increasing interest of European powers in the Middle East led local and later colonial officials to develop laws to control excavation and exportation of antiquities. These laws reflect tensions between nations over ownership of the past. Were historical artifacts from the Middle East more meaningful to people who live in the region, or do they reflect a world historical tradition shared by all nations?

In the Middle East, these tensions increased in the aftermath of World War I, as Western scholars such as James Henry Breasted developed the idea that Western civilization truly began in the ancient Middle East, and as nationalist movements in the region increasingly asserted their own close connection to the historical traditions of their homeland. One manifestation of this rise in nationalism was the move by many nations to enact national ownership laws, which vested ownership of antiquities in a nation and created a set of rights that were recognized when antiquities were removed. These issues are if anything even more contentious today, as they continue to be debated in the global community of the twenty-first century. Thus the history of antiquities law in the Middle East is a history of attempts to control the past by owning its most tangible remains.

ANTIQUITIES LAW IN THE OTTOMAN EMPIRE

In response to increasing foreign interest in parts of the Middle East and the looting of archaeological material from the Ottoman Empire, an Ottoman Antiquities Law was passed in 1874 for the regulation of the movement of antiquities uncovered during archaeological excavations. This early antiquities law recognized the right to a division of artifacts (between landowners, the empire, and foreign archaeological teams), although the ownership of the cultural heritage was vested in the empire.

A subsequent Ottoman law passed in 1884 established national ownership over all artifacts in the Ottoman Empire and sought to regulate scientific access to antiquities and sites. Under the 1884 law, all artifacts discovered in excavations were the property of the Imperial Museum in Constantinople and were to be sent there until the Director of the Museum made decisions about the partition of the finds. This law, although viewed as a national ownership law, alternatively could be interpreted as legalized cultural imperialism — the Ottoman Empire preserving not only the archaeological legacy of its core but also appropriating material from its territories in the periphery. By controlling archaeological goods, the Ottomans effectively regulated European access to its heritage, access that had previously gone unchecked.

The Ottoman Antiquities Law of 1884 followed closely on the heels of several major European expropriations from the Turkish heartland, including that of the Pergamon Altar (now in Berlin). The 1884 law did not develop in response to local Turkish interest in cultural heritage, but rather as a measure to ensure that artifacts from the far-reaching Ottoman Empire remained within its boundaries. The primary drafter of this legislation, Osman Hamdi Bey, Director of the Imperial Museum, was concerned with filling the coffers of the museum with the splendors of the empire, sending a clear political message to the West, and potentially capitalizing on tourism to the area.

A national stake in the cultural heritage of the empire was established as Chapter I Article 3 of the 1884 law: "all types of antiquities extant or found, or appearing in the course of excavation automatically belong to the state and their removal or destruction is illegal." The nationalization of cultural artifacts and heritage was a reaction to the imperialism of the eighteenth and nineteenth centuries when the subjects and citizens of Western empires ransacked the monuments of less well-developed nations. The combination of the declining Ottoman Empire and Western expansion was the major impetus behind the inclusion of a national ownership element in the law. Although cultural heritage was the property of the empire, local Turks could still buy, sell, and exchange artifacts within the territorial boundaries of the empire.

Under the 1884 law, all foreign excavators in the Ottoman Empire had to apply for permission to excavate and all antiquities recovered during excavation were to be transferred to the Imperial Museum in Constantinople (Chapter I Article 12). The Director of the Imperial Museum was to make all decisions concerning the disposition of artifacts and whether redundant material was available for export from within the borders of the empire. Once artifacts were determined to be non-essential to the cultural heritage of the Ottoman Empire, they were returned to the excavator and/or landowner for study, analysis, or sale. Excavators were only allowed to take casts and molds of the exceptional artifacts or, if the material was considered redundant or unnecessary for the museum collection, it could be exported. According to Shaw (*Possessors and Possessed*, p. 116), "efforts to control the apportionment of antiquities to the Imperial Museum and to foreign museums often depended on the latter's interest in promoting Ottoman cultural aspirations," thus using the preferential export of artifacts as indicators of power and influence in the political arena. The Ottoman government also deployed the artifacts as cultural capital in the form of bribes and gifts in order to forge political alliances and cultivate diplomatic relationships. At the end of the nineteenth century, the close relationship between the Sultan Abdulhamid II, Kaiser Wilhelm I of Prussia, and Emperor Franz Joseph I of Austria was enhanced by bypassing the law when it suited diplomatic goals, allowing the free movement of archaeological material for political goodwill (Shaw, *Possessors and Possessed*, p. 117).

In practice, enforcement of this law was virtually impossible. The expanse of the empire was so great that the Ottoman government did not have enough officials to oversee and implement the various regulations of the 1884 law and the inherent bureaucracy of the empire delayed excavation permits for almost a year. Excavators who previously had unregulated access to the finds from their forays into the field were extremely unhappy with the new provision that all finds had to be vetted by the Imperial Museum prior to study, analysis, and/or export.

In an effort to curb the loss of cultural heritage from the empire, Chapter I Article 8 of the 1884 law specifically stated that without the permission of the Imperial Museum the exportation of artifacts was prohibited. Many foreign archaeological missions and locals contravened the law almost immediately after its enactment, including the University of Chicago Expedition to Bismaya led by Edgar J. Banks (Wilson, *Bismaya*). An intricate smuggling network developed through the region. Public awareness of artifacts as commodities and consumer demand played integral roles in the legal and illegal movement of artifacts. Under the 1884 law artifacts accidentally recovered by a landowner entitled him to a share of the artifacts' value. Thus the legislation imbued archaeological artifacts with an economic value that may not have previously existed. Those who recovered artifacts received financial compensation prescribed by law — the looter was rewarded. The general feeling among archaeologists was that the 1884 Ottoman Antiquities Law, sound in principle, practically gave free hand to the plunderers of ancient remains while at the same time placing serious impediments in the way of legitimate excavators. The Ottoman Antiquities Law of 1906 was passed to strengthen the 1884 law and to close many of the loopholes, but there was still much movement of material within the empire and disgruntled foreign excavators lamented a lack of access to their archaeological material. In the waning days of the empire these initial attempts at regulating the movement and distribution of the cultural heritage of the region formed the basis for much of the later and currently existing legislation in the region.

EGYPT

Although Egypt was an Ottoman province it was difficult to rule from such a distance. As such it remained semi-autonomous, under the jurisdiction of a local khedive and often acted autonomously until it was conquered by Napoleon's troops in 1798. This resulted in the large-scale removal of various monumental Egyptian sculptures, which made their way to the British Museum and into the consciousness of the West. This led to an ongoing fascination with Egypt and the Middle East and to calls by Egyptian academics for the protection of Egyptian cultural heritage. Later acquisitions by the consul collectors, which included British Consul General Henry Salt and his French rival Bernardino Drovetti, did not go unnoticed by Egyptian scholars even though the removal of artifacts was undertaken with *firmans* (permission) issued by the local khedive (Jasanoff, *Edge of Empire*, p. 230).

In 1835 Muhammad Ali, then Ottoman khedive of Egypt, issued an ordinance forbidding the export of antiquities, establishing an Egyptian Antiquities Service, and proposing the establishment of a national museum. One of the first of its kind, the ordinance laid the blame for the despoliation of the country squarely on the West, at the same time citing contemporaneous European legislative efforts setting a precedent for protections against the movement of antiquities (Reid, *Whose Pharaohs?*, p. 93). In 1834 Greece had enacted a national antiquities law that protected against the illegal export of archaeological material from Greece without the permission of the Greek government. But Egypt was not in Europe and was the target of European domination, which included the appropriation of archaeological artifacts for European institutions. Various diplomats to Egypt viewed the ordinance as an attempt to monopolize the past — an Egyptian attempt to control the nation's history, which some argued was the history of the world. The West was besotted with Egyptian artifacts and the demand for such items encouraged foreign missions to scorn the export ban (Reid, *Whose Pharaohs?*, p. 57).

Artifacts could still move freely around the Ottoman Empire, often making their way outside the empire through the porous border and with the aid of diplomatic cover. The Egyptian governor frequently used artifacts as cultural capital, viewing them as bargaining chips to be exchanged for European diplomatic and technical support (Reid, *Whose Pharaohs?*, p. 54). They were also used as gifts for the sultan in Istanbul, once again acting as material ambassadors, engendering goodwill. In 1882, after increasing instability at the end of the nineteenth century, Britain invaded the region to protect their interests in the area. Increasing curiosity in archaeological excavations and Egyptomania may also have played a role in Britain's desire to control the region.

As early as 1880, Egypt enacted a national ownership statute, a decree from Khedive Mohamed Tewfik, which clearly stated, "all the monuments and objects of antiquity, recognized as such the Regulation governing the matter, shall be declared the property of the Public Domain of the State."

In 1912, Egyptian Law of Antiquities No. 14 was passed clearly as a result of the earlier Ottoman efforts as well as the 1880 Egyptian Antiquities Decree. A national ownership law, Article 1, vested the title to both excavated antiquities and those yet to be discovered in the national government. The Minister of Public Works, placing antiquities in Egypt squarely in the sphere of the built environment, issued the permits. There was a provision outlining the division of finds, which were to be divided equally into two shares — one for the state and one for the excavator. The law expressly stated that the division was to be made by the Antiquities Service, but the excavator had the right to choose his portion (Article 11). It also provided for the acquisition of objects of national importance by the State, after fair compensation to the excavator. This law acknowledged the Western fascination for all things Egyptian.

Article 13 of the 1912 law outlined a provision for the licensed sale of antiquities. With the authorization of the Antiquities Service, merchants were granted permission to buy and sell antiquities. Artifacts could then be exported with the proper permits. The entire process was under the oversight of the Minister of Public Works, thereby ensuring that only select pieces left Egypt. It was under this law that Breasted (in 1925) acquired the Senenmut statue for the Field Museum: "I convinced the dealer that the statue could be shown to the Museum in Cairo and legal arrangements made for its export without a risk of it being seized by the government ... the Government official in charge of the Department of Antiquities hesitated some time before permitting its export, but to make a long story short, I now have the piece in my possession and all arrangements for its legal export have been made" (*JHB to Mr. Stanley Field, President of the Board of Trustees, Field Museum, April 4, 1925*).

POST-WORLD WAR I LAW

With the demise of the Ottoman Empire, newly independent areas including Palestine and Iraq began to enact their own forms of national legislation, beginning in both cases with efforts by colonial officials to draft antiquities laws and build national museums, and furthered by local nationalists. In order to maintain a sense of normalcy and to continue legal oversight, these newly formed independent states often adopted similar laws, especially when they were deemed reasonable (Chatterjee, *The Nation and Its Fragments*). Typically, countries did not act in isolation when attempting to protect their pasts, but rather learned from each other's successes and failures, with the common goal of retaining their cultural heritage as a signifier of a rich past and a bright future. In this case, the Egyptians looked to Greece and Turkey; Iran and Iraq took note of the rampant export of monumental Egyptian sculpture and architectural items and based their legislation on Egyptian efforts. Middle Eastern antiquities laws often possessed common elements: most instituted a system of partage, whereby archaeological material was divided between the excavator, the state, and the landowner. This system of dividing the spoils attempted to ensure that the finest elements and representations of a nation's cultural heritage remained within its boundaries. Early laws established repositories for the various artifacts in the form of national museums. In some instances this required a separate law, but these repositories were intended to coincide with the partage system, so that there would be no accusations of a country's inability to care for these manifestations of their heritage. A department of antiquities was also created as the mechanism to oversee both the national museum and to enforce the facets of these laws. There was an ideological divide as to whether these agencies were located within the Ministry of Education or the Ministry of Public Works, each signifying a different importance for archaeological artifacts — as educational tools in building a national identity or as part of the built environment to be preserved and protected. Most of the early ordinances and laws regulated the exportation of antiquities outside their territorial boundaries. Many of these initial legal proscriptions continued and can be found in the current legislative efforts of these nations.

PALESTINE

On December 9, 1917, in the final stages of World War I, Jerusalem (then under Ottoman rule) surrendered to the British forces commanded by General Allenby. This act marked the end of four centuries of Ottoman domination and the beginning of thirty years of British rule in Palestine. Using the 1884 Ottoman law as a springboard, the British authorities charged with cultural heritage protection in the region promulgated the 1918 Antiquities Proclamation, which noted the importance of cultural heritage. In July 1920, the Mandate civil administration took over from the military, and archaeologist John Garstang was appointed as the Director of the Department of Antiquities for Palestine. In a report of his activities to the Palestine Exploration Fund, Garstang ("Eighteen Months Work," p. 58) stated, "the Antiquities Ordinance was based not only on the collective advice of archaeological and legal specialists, but embodied the results of experience in neighbouring countries." Garstang established an Antiquities Ordinance for Mandate Palestine vesting the ownership of moveable and immoveable cultural heritage in the Civil Government of Palestine (that is, the local indigenous government rather than the foreign occupiers). Vesting national ownership of cultural material of the state in the civil government and not the occupiers was a departure from examples of earlier antiquities law, which gave those in charge (the Ottomans) the power to make decisions over the disposition of the cultural heritage of the region. The enactment of this ordinance established a Department of Antiquities and an archaeological advisory board (comprised of representatives of the various foreign archaeological schools in the area), ensuring that the protection and oversight of the cultural heritage in Palestine was carried out locally rather than from an imperial capital. In reviewing the legislative developments of an ordinance to protect archaeological heritage in Palestine it seems that without the direction and impetus of the Mandate government, specifically the guidance of John Garstang, such laws would not have been achievable by the divided local population.

The primary goal of the ordinance was the protection of archaeological antiquities and sites. The regulation of ongoing archaeological excavations was monitored by the Department of Antiquities, as was the sale of artifacts. In response to criticisms of the earlier 1884 law by archaeologists and tourists regarding the lack of access to archaeological material, a provision was included for the sale of material deemed unnecessary for the national repository, a decision made by the Director of Antiquities and the Board of Advisors. The Department was given permission by the High Commission to issue licenses for the trade in antiquities. In 1920, for the first time in Palestine, a licensed trade in antiquities was regulated and overseen by a bureaucratic entity — the Department of Antiquities. Article 21 of the Palestine Mandate of the League of Nations of 1922 further cemented the right to scientific access for nationals and foreigners by ensuring access to excavations and archaeological research for any member of the League of Nations. Scientific archaeological enquiry and the distribution of archaeological material took center stage during the Mandate period, embodied in Antiquities Ordinance No. 51 of 1929.

IRAQ

During the Mandate period, the British government also received a mandate from the League of Nations to oversee the political development of what is now Iraq. The Mandate oversight was established to prepare for Iraq's impending independence, but with British interests in mind. An element of Britain's interests included a continuation of access to archaeological sites and artifacts. As part of the Ottoman Empire, the region of present-day Iraq had been subject to the antiquities laws of 1874 and 1884 and foreign archaeologists in Mesopotamia had been acting both within and outside the regulations (Bernhardsson, *Reclaiming a Plundered Past*). Prior to the Mandate period, artifacts from this region were considered international and moved without restriction to collections in the Middle East, Europe, and North America. Although not ratified, the Treaty of Peace with Turkey, commonly known as the Sèvres Treaty, became binding between Britain and Iraq under a 1922 treaty, which obliged Iraq to adopt the provisions in the Sèvres Treaty. Under these requirements excavations were subject to government approval. Also included were measures to protect foreign missions from discrimination. In light of the new provisions, which included assurances of partage, Iraqi scholars questioned the objectivity of the British government in drafting new antiquities legislation.

Formalized antiquities protection was in a state of limbo when Breasted visited in 1919–1920, but both the British and Iraqis were moving toward hybrid legislation, with Briton Gertrude Bell helping to formulate antiquities legislation for Iraq. The proposed legislation would still allow foreigners like Breasted to access sites and artifacts, but it also took into account Iraqi concerns. Resistance to Bell's drafts of antiquities legislation led to a confrontation with Sati' al-Husri, Director General of Education in Iraq (1921–1927) — as a result, archaeology in Iraq came under the purview of the Department of Education. Al-Husri's major disagreement was over a national ownership clause, and he could point to examples of national ownership antiquities laws in Greece, where everything went to the national museum. Bell argued that if all things belonged to the state, no foreign archaeologist would come to Iraq to excavate (Bernhardsson, *Reclaiming a Plundered Past*). This battle would continue even though Bell's antiquities law eventually passed in 1924.

The law contained many of the same provisions and characteristics mentioned previously, with one exception. Rather than the equal division of artifacts (partage) between the national museum and the excavator or landowner, Article 22 of the new law stated:

> At the close of excavations, the Director shall choose such objects from among those found as are in his opinion needed for the scientific completeness of the Iraq Museum. After separating these objects, the Director will assign to the [excavator] such objects as will reward him adequately aiming as far as possible at giving such person a representative share of the whole result of excavations made by him.

The law left the division of finds to the discretion of the director of the Iraq Museum and attempted to ensure that the pieces of greatest importance remained in Iraq.

LEGAL CHRONOLOGY IN ANTIQUITIES LEGISLATION

1835	Antiquities Ordinance (Muhammad Ali) (Egypt)
1834	National Antiquities Law (Greece)
1874	Ottoman Antiquities Law
1880	Decree on the Prohibition of the Export of Antiquities (Egypt)
1884	Ottoman Antiquities Law
1906	Ottoman Antiquities Law
1912	Law of Antiquities No. 14 (Egypt)
1918	Antiquities Proclamation (Palestine Mandate)
1920	Treaty of Peace with Turkey (Sèvres Treaty)
1922	The Palestine Mandate of the League of Nations
1924	Antiquities Law (Gertrude Bell) (Iraq)
1929	Antiquities Ordinance No. 51 (Palestine)
1930	Antiquities Rules, February 1, 1930 (Palestine)
1934	Antiquities Ordinance (Palestine)
1936	Antiquities Law No. 59 (Iraq)
1948	Establishment of the State of Israel (reinstitutes the AO 1929)
	The Gaza Strip (governed by Egypt and the AO 1929)
	West Bank (governed by Jordan and the AO 1929)

The post-World War I period was a mix of old and new as antiquities legislation in the newly forming states struggled to encompass the desires of the nationalists and the foreign supporters of archaeology. This was the legislative quagmire in which Breasted would find himself when he next returned to the Middle East.

CONCLUSION

At the end of formal colonialism, national laws and cultures in the former colonies were not only modes of resistance, but were proof of colonialism's perpetual victory over the colonized. For many the fact that the Parthenon marbles or the bust of Nefertiti rest in Western institutions is a sign of the perpetuation of colonialism. Many today view the great museums of the world as testaments to imperialism and colonialism, while others see these universal museums as proof of the ingenuity of mankind — artifacts as effective ambassadors showcasing the wonders of other cultures.

6. THE ARAB REVIVAL, ARCHAEOLOGY, AND ANCIENT MIDDLE EASTERN HISTORY*

ORIT BASHKIN
UNIVERSITY OF CHICAGO

When James Henry Breasted toured the Middle East in 1919–1920 he saw not only objects, but also people, namely the Arab inhabitants of the region. For their part, the Arabs themselves, or more specifically, educated Arab elites in the Levant and Egypt, became increasingly interested in the themes and questions to which Breasted devoted his works: archaeology, the history of the ancient Middle East, and the connections between Middle Eastern and Western cultures and between history, archaeology, and civilization. To these Arab elites, knowledge of world history in general, and of the ancient Middle East in particular, came to be understood as the signifiers of modernity itself. In this realm, archaeology as a science, and archaeologists as its practitioners, were not scorned as representatives of Western colonialism and imperialism, but rather were thought of as integral members of the modern scientific community. Moreover, since the late nineteenth century, Arab literati and intellectuals emphasized (like Western archaeologists) the fact that the people of the Middle East had contributed much to world civilization, underscoring the need to learn more about the ancient Middle East and its connections to the foundations of modern civilization. This essay looks at the reception of Breasted's popular work *Ancient Times* in the Middle East and explores the ways in which Arab intellectuals investigated themes related to Middle Eastern archaeology, linguistics, and ancient history.

In recent years, historians of Arab nationalism have looked at the ways in which the rise of archaeology as a scientific discipline served modern Arab nation-states. These historians have studied not only the connections between European and American archaeologists and the colonial powers, but also, and more significantly, the ways in which Arab states appropriated the science of archaeology to support a host of national projects in the 1920s and 1930s and beyond. Moreover, modern nation-states such as Egypt, Iraq, Syria, and Lebanon have used the ancient past to support narratives of ancient golden ages and to sustain various territorial and political claims.

As shown by Wendy Shaw in *Possessors and Possessed* (with respect to the Ottoman imperial center) and Donald Reid in *Whose Pharaohs?* (with respect to nineteenth-century Egypt), there was a great deal of interest in archaeology and ancient history in both Istanbul and Cairo before the collapse of the Ottoman Empire. In other Arab provinces of the Ottoman Empire, the main venue for exploring the ancient past was the nascent Arab print media, in particular the press. After World War I, with the formation of Arab nation-states, narratives about Egyptian, Mesopotamian, and Phoenician cultures permeated the print media, public spheres, and national cultures that emerged in the Middle East. These efforts, moreover, were popularized in the new states' textbooks and national museums. The discovery of Tutankhamun's tomb in 1922 influenced Egyptian architecture that attempted to incorporate ancient Egyptian styles into the new state's monuments, inspired new works of prose and poetry that hailed the culture of ancient Egypt, and led to the rise of a new generation of Egyptian archaeologists. In Lebanon, Christian Arabs championed the culture of the Phoenicians and represented them as the forefathers of Western civilizations. As shown by Ernest Dawn, scholars, intellectuals, and educators in Iraq, Trans-Jordan, Syria, and Palestine labored to underline the connections between the cultures and histories of the Babylonians, Assyrians, Nabateans, and Arabs in order to construct the world of the Semites, an ethnic group that had bestowed upon the world its first civilizations, urban centers, and writing systems. These Middle Eastern intellectuals contended that the formation of the first Islamic state did not mark the beginning of the history of the Arab nation and identified the empires that established the foundations of Arab civilization as the "Semitic Empires."

---

* I thank Geoff Emberling and Emily Teeter for their kind assistance

ARABIC *ANCIENT TIMES*

Breasted's own works played an immensely important role in the formation of modern identities. In 1926 *Ancient Times* was translated into Arabic (*al-'Usur al-qadima*) by Da'ud Qurban, and a second edition appeared in 1930 (fig. 6.1). Like the English original, the translation was intended to serve as a textbook for high-school students. Thus, akin to the original, each chapter ended with study questions to help prepare students for exams and teachers for their lessons. The project turned out to be extremely successful. According to the introduction to the second edition, the Arabic rendition won praises in the local Arab press and the ministries of education in various Arab countries showed interest in the book, which was taught in schools in Palestine, Iraq, and Trans-Jordan in the interwar period. The success of the book should not come as a surprise. *Ancient Times*, which related the history of ancient Middle Eastern empires to the West and distinguished between Semites and non-Semites (although for purposes that differed sharply from those of the Arab intellectuals who were taken with the book) corresponded to the desire of Arab nationalists to illustrate that the ancient Middle East had helped shape the civilization of the modern world. Furthermore, many of the Arab intellectuals belonged to a Westernized elite class who believed that Arabs, while engaged in the battle against colonialism, should also integrate certain Western elements into their social, cultural, and political life in order to modernize and progress. To justify their claims historically, many emphasized the connections between the East and the West, and the influence of Middle Eastern cultures (e.g., civilization of ancient Egypt or the translations of Greek masterworks into Arabic) on the West. They looked to models of hybridity and reciprocity in the past as a way of validating cultural choices made in the present. *Ancient Times*, which attempted to connect the history of the Middle East with that of Greece and Rome, fit their goals.

The book was at times abridged and modified. I will not dwell here on the sections of text that were omitted, but rather will make a few general comments about the functions of the text within an Arab national context. Harold Nelson, who played a key role in the production of the Arabic rendition, added his own introduction in Arabic in the hope of addressing some of the concerns of the Muslim and Arab readers. Nelson characterized Breasted as "the greatest American historian who writes on the history of the East today." The book, he noted, was meant to foster pride among the Arabs concerning their history and appreciation of the valuable role their forefathers had played in the formation of science and civilization:

> Some of the extremists amongst our brethren in Muslim lands tend to consider all that had preceded the arrival of the prophet Muhammad as if it belongs to a different world.... Clearly, this is destructive to the understanding of the history of the Islamic world, because civilization and culture did not start from the time of the Prophet....
>
> The blood of those who were under the rule of the Pharaohs, the kings of Babylon, the Assyrian Empire, the state of the Hittites, and the leaders of the civilized settlements in Palestine and Syria, runs in the veins of those who live in these countries [today].... Their ethics, the common practices apparent in their lives, and their philosophy originated from these ancient times.... We argue ... that if the denizens of the East today stayed in the cities of Assyria or Phoenicia or Egypt, which existed 4000 years ago, they would feel that they were in their homeland (from Harold Nelson's Introduction in J. Breasted, *al-'Usur al-qadima*, Second edition, 1930, p. ii).

Nelson here critiques the Islamic position that regarded the period before the arrival of the Prophet Muhammad as the period of ignorance, marked by the Arabic word *jahiliyya* and characterized by idol worship and wrongdoing. Nelson's secular assumptions that the histories of the Middle East did not begin with the rise of Islam were also accepted by Arab thinkers, who, at the time, nationalized the Islamic past and tied it to the history of the region in antiquity. Moreover, his introduction was included in a textbook that was recommended by the ministries of education in various Arab countries.

The secularizing nature of the text has also to do with the act of translation, as the book secularized Islamic terms in order to give meaning to the terminology used by Breasted. This process was well underway by the mid-nineteenth century and was not unique to the translation of *Ancient Times*, but it is keenly felt in *al-'Usur al-qadima*. Thus, the word *umma* did not signify the Islamic community of believers, but rather served as the translation of "nation" in Breasted's usage. Similarly, Breasted's chapter, "Men before Civilization," was translated as "The

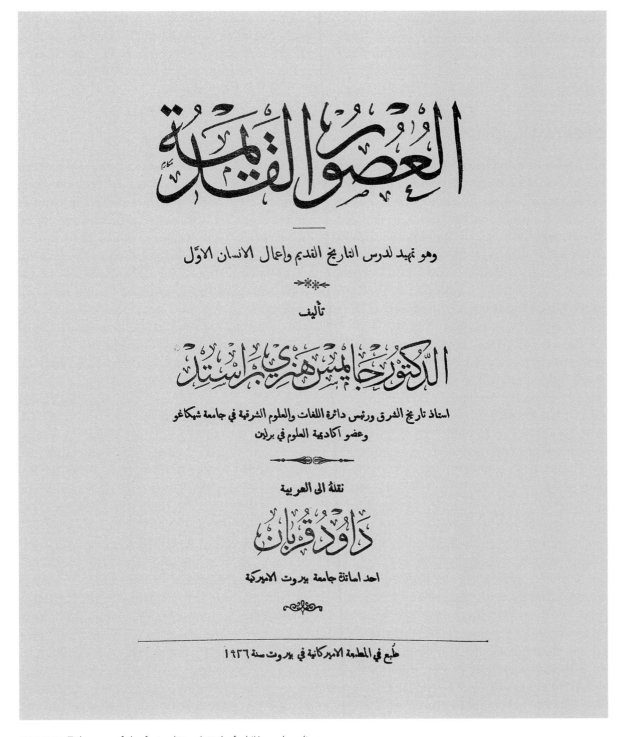

FIGURE 6.1 Title page of the first edition (1926) of *al-'Usur al-qadima,*
Da'ud Qurban's Arabic translation of J. H. Breasted's *Ancient Times*

Jahiliyya of Ancient Men [*aqdamiyyin*] and their Gradual Progress." *Jahiliyya* was thus transformed to signify the period before the rise of civilization. When Breasted used the word "forefathers" it was translated as *aslaf*, a word used in the Islamic tradition to signify the pious ancestors [*salaf*] of the patristic period of Islam. The connections between ancient history and the modern world in which students lived also affected the translation of the questions which ended each chapter. Some of Breasted's questions were changed to better fit the context of the Arab student, as students were asked to write what archaeological evidence might be found near to where they lived. As part of the description of a particular type of tent in the chapter "Western Asia: The Land of Babylon," students were told that these were the same as "the tents of the Bedouin in our days" and that "as we divide today the tribes into nomads and civilized so did our forefathers." This comment is important not only in reference to Breasted's original, but also because it reflects the assumption of the translator that students might have actually seen Bedouin and their dwellings and would consequently find these descriptions somewhat familiar.

ARABIC REVIVAL AND ARCHAEOLOGY

Local elites and governments, especially in Istanbul and Cairo, had already begun to take a serious interest in archaeology and ancient history by the time Breasted visited the region. Egyptian elites in particular had invested some effort in guarding the ancient treasures of their country in terms of legislation and the construction of museums, as noted elsewhere in this volume. Furthermore, important trends in the Arab world that had arisen in the mid-nineteenth century were helping to spark the curiosity of the Arab intelligentsia by 1920, with regard to the region's ancient history. One of these trends was the Arab cultural revival that began in the late nineteenth century, signified by the Arabic word *al-nahda* in Lebanon, Syria, and Egypt. The *nahda* entailed the renewal of Arabic literature and culture, manifested especially in projects of translations of European works into Arabic, and the adoption of new modes of writing, such as the newspaper article, the novel, and the short story, as vehicles of literary and cultural expression. The *nahda* also brought about a re-evaluation of the relationships between East and West, with a call for greater openness to Western ideas (ranging from political theory to styles of dress) and the usage of these very same categories (East/West) as relevant to the identities of modern Arabs. With the *nahda* emerged the modern Arabic print media, and literary and cultural journals enjoyed great popularity in the region. Many of these journals published articles on the kingdoms of Assyria, Babylonia, and Egypt as well as on the histories of the kingdoms of Judah and Israel. Most of the journals originated in Lebanon; their editors were graduates of the newly established Syrian Protestant College (later the American University of Beirut) who were in close contact with the American missionaries who worked there (some of whom had been trained in Semitic philology). The missionaries' major project, the translation of the Bible into Arabic (by the Protestants in 1856–1865, and by the Catholics in 1876–1880), in which some of the Arab Christian intellectuals were involved, led to debates about the origins of Semitic languages and the relationships between them. Lebanese and Syrian intellectuals therefore expressed their opinions on the subject in the pages of the press. The European and American archaeological excavations in the region had likewise kindled the interest of Arab intellectuals, who discussed the conflicts between archaeological evidence and the Bible and the Quran in the journals of the time.

Beginning in the last part of the nineteenth century, the movement of Arabism did much to shape the lives of Arab literati. This nascent form of Arab nationalism was rather nebulous in comparison to its articulations in the twentieth century. We know today, for example, that many Arabs supported the Ottoman side during World War I and preferred the existence of a Muslim Empire to Western colonialism and imperialism. Nonetheless, great attention to Arabic language and literature and Arab culture was noticeable in the print media in Egypt and the Fertile Crescent, especially in Beirut and Damascus. Moreover, in Egypt a fascination with ancient Egyptian culture took hold among the intelligentsia. This interest began in the mid-nineteenth century and continued into the interwar period. As Donald Reid, Israel Gershoni, and James Jankowski have shown, Egyptian nationalists like Ahmad Lutfi al-Sayyid, Muhammad Husayn Haykal, Salama Musa, Tawfiq al-Hakim, and Taha Husayn underlined the connectivities between the ancient Egyptian past and the modern Egyptian present, took great pride in the achievements of the ancient Egyptians, and argued that the culture of ancient Egypt, as formed in the Nile Valley, gave Egypt a unique identity that distinguished it from the rest of the Arab Middle East. The sense that Arabs and Egyptians were different from Turks, despite the fact that all were Muslims, became increasingly salient. This sense of difference came to be felt by Christian Arabs as well, who held that language, history, and a shared culture rather than religion should be the new parameters by which an individual's belonging to a particular community should be determined. The new debates about the meaning of being a modern Arab led to questions regarding origins. When Arabs asked who they were, in other words, they also asked where they had come from.

ARABIC-LANGUAGE JOURNALS

To ascertain what educated Arab elites knew about the ancient Middle East when Breasted set off on his journey, let us turn to the two cultural and scientific journals that were widely read in this period, *al-Muqtataf* and *al-Hilal*, as well as to a publication of the Arab Language Academy (*al-Majma al-ilmi al-arabi*) which had been started in Syria. *Al-Muqtataf* and *al-Hilal* were the most prominent journals dealing with popular science, history, and archaeology in the Arab East. They were highly prestigious publications, the editors of which nurtured the Arab intellectuals most closely affiliated with the *nahda*. The journals were trans-regional publications, and were read by Syrians, Egyptians, Palestinians, Iraqis, and other educated Arabs in the Middle East, Europe, and the Americas. Most of the journals' readers, as well as their writers, were men, but we do find letters and columns written by women as well. Despite the high illiteracy rates in the Middle East, scholars of modern Arab culture concur that these journals were one of the major platforms that ignited the revival of modern Arab culture and history.

Al-Muqtataf was started in 1876 as an initiative of two Lebanese intellectuals, Yaqub Sarruf (1852–1927) and Faris Nimr (1852–1951), and *al-Hilal* was started in 1892 by Jurji Zaydan (1861–1914). Zaydan, as a historian who claimed to know several Semitic languages, was very interested in archaeology and referenced archaeological evidence in his books about the history of the Islamic East. He and other contributors to these journals also wrote historical novels whose plots were sometimes grounded in the periods of antiquity and late antiquity. Sarruf, Nimr, and Zaydan all moved to Egypt in the 1890s and despite being Lebanese saw their journals rise to prominence in the Egyptian public sphere. Some of the intellectuals who wrote in *al-Muqtataf* and *al-Hilal* were characterized by noted historian Albert Hourani as "Christian secularists" because of their firm belief in the power of science. In fact, they clashed with their missionary sponsors in the Syrian Protestant College over their conviction that Darwinism should be introduced to the students of the college. Their journals, however, were also read by Muslims and Jews all across the region. The articles published in these journals give us a sense of what educated people, graduates of the new high schools and academies in the region, read and discussed. These journals, moreover, often mention a number of scientific and cultural societies and literary salons in which matters related to archaeology and ancient history were debated.

The journal of the Arab Language Academy, *Majallat al-majma al-ilmi al-arabi*, was a publication of a different sort. The academy was established in Syria in 1918 through the initiative of the Syrian Muslim intellectual and literary critic Muhammad Kurd Ali (1876–1953). Its members were mostly renowned Muslim scholars, and its journal featured more scientific articles (as opposed to *al-Muqtataf* and *al-Hilal*, which tended toward popular history and science). The academy held lectures and meetings for its members starting in 1919, supported a publishing house (established in 1920) and a museum, and began publishing its journal in January 1921. Its members gradually came to include the elite of the Arab cultural sphere: historians, poets, educators, and intellectuals from Syria, Lebanon, Egypt, Iraq, Palestine, and North Africa. Although in terms of its character this publication differed markedly from *al-Muqtataf* and *al-Hilal*, its authors were interested in similar themes and devoted much attention to linguistic and cultural issues pertaining to ancient Middle Eastern history and philology.

In *al-Muqtataf* and *al-Hilal*, stories about archaeology and ancient Middle Eastern history appeared in a few modes: articles devoted to a particular historical or archaeological theme, sections dedicated to scientific discoveries, and questions addressed to the editor from readers who were curious about the local histories of the cities and provinces in which they lived. Readers of these journals during the years 1919–1921 thus learned about new theories regarding the construction of the pyramids, ancient Egyptian burial practices, and everyday life in ancient Egypt. One story about the history of Iraqi Jews, written by the Iraqi Christian intellectual Yusuf Ghanima, included data about Ur and the excavations carried out in the region by British consul J. E. Taylor. Stories on archaeology were not limited to finds related to the ancient Middle East, as the period of late antiquity was also of importance: one piece reported on the excavations in Ashkelon and the discovery there of a church from the fourth century AD.

Ancient Greek history and archaeology were important because some Arab authors strongly believed in the significant connections between the Greeks and the Phoenicians, as well as between the Greeks and the ancient Egyptians. This was the basis for these authors' claim that the peoples of the Middle East had a crucial role in the formation of Western civilization. For the most part, though, pieces in *al-Hilal* and *al-Muqtataf* about ancient Greece referred to these connections only occasionally, and items seemed to have been published primarily because editors sensed that the readers would find them fascinating. Items and stories along these lines appeared that discussed Minoan civilization and the ruins at Knossos, the statue of Venus de Milo and its discovery in 1820 and the debates among archaeologists about its origins, and Greek mythology. Moreover, the journals also published articles about excavations in Mexico, America, and other parts of the world in order to establish the validity of archaeology as a means of elucidating the history of ancient continents and civilizations. In January 1920, *al-Muqtataf* ran a story on how archaeology could assist the discovery of the bones of dinosaurs and the development of the science of natural history. The coverage of global ancient history was linked to the desire of many writers to position their own history within that of the world at large. One article in *al-Hilal*, for instance, chronicled the movement of civilization and progress from south to north, in an attempt to explicate why the cultures of China, India, and the Middle East did not play a major role in the contemporary world. Readers also directed questions to the editor about the Iron Age and findings from this period.

Archaeology, then, was needed in order to add to knowledge about the Middle East and to complete the educated individual, who was well informed about recent scientific discoveries and the history of the globe. Most importantly, archaeology was presented as a science. The increasing usage of the word "evidence" [*adilla*] illustrates this point. Many articles on historical themes refer to available archaeological evidence to corroborate information or a thesis. An essay in *al-Muqtataf* entitled "The Semitic Inhabitants of Syria" states this clearly:

> No other type of evidence supports the construction of history [*bina' al-ta'rikh*] as archaeological findings and discoveries hidden in the depths of the earth.... For this reason, ancient history is always subjected to correction and renewal, based on new findings and excavations, and it correlates to the state of the country, concerning excavations, and the country's own progress or decline (*al-Muqtataf*, August 1920: 128–29).

Similarly, the Arab Language Academy viewed archaeology as an important science and saw that it could be useful in the reconstruction of Islamic history and the development of the Arabic language. It supported two committees as part of its work, a language and literature committee and a scientific committee comprised of archaeologists who both collected and curated displays in the academy's museum. The first issue of its journal featured a long essay detailing all the finds available in the museum of the academy.

Furthermore, it was understood that archaeology was not mere authentication of information conveyed in the Bible or the Quran (even though articles referenced Western biblical archaeologists). Many readers and writers did write about the harmonious relationship between archaeological findings and the Bible or the Islamic tradition. Others, however, considered the Bible as just one of a number of sources that could tell readers something about the history of the region, their own religious views notwithstanding. For instance, *al-Muqtataf*'s piece from August 1920 on the history of Syria from the second millennium BC to the Islamic period used as sources "archaeological findings [*athar*], inscriptions that are found in this region," as well as the Torah and written texts by Egyptians and Assyrians to speculate on the demography of Syria, its relationships with other kingdoms in the Middle East, the languages spoken in Syria, and the nature of its civilizations. Articles on the kingdom of Lydia, published in the same issue, compared information from the book of Genesis 10:21, from Josephus Flavius (AD 37–ca. 100), and from the writings of Arab chroniclers and geographers such as Ibn Khaldun (1332–1406), Ibn Ishaq (d. 767 or 761), and Ibn Hazm (994–1064) and their understanding concerning the traditions about Lud, son of Shem.

These articles taken together communicate the sense that the origins of the information about the ancient Middle East were not a matter of much concern. Therefore, regardless of whether it was a European or American historian, philologist, or archaeologist who provided valuable insights about the Middle East, Arabs needed to be aware of the most up-to-date information. An *al-Hilal* article on the origins of the word "*al-'iraq*" by Iraqi writer Razuk Isa cited a few Muslim scholars including Yaqut al-Hamawi (1179–1229) and al-Asmai (ca. 740–828), hypotheses suggested by Guy Le Strange, and words in Akkadian and Sumerian only to come to the conclusion that "the meaning of the word '*Iraq*' still remains a hidden secret" which none of the Orientalists had managed to unlock. Isa ended the article by saying:

> I ask our Orientalists [*'ulama'una al-mustashriqun*]: What is the meaning of '*iraq* in your opinion? Have you found anything in your studies about it in Babylonian or Assyrian inscriptions? In a book or a chronicle where the name of '*iraq* was first mentioned? Did these lands know the name of '*iraq* before the Persian conquest?" (*al-Hilal* 28 [1919–1920]: 737–41).

Writers on Islamic themes also looked at archaeology as a science that could tell much about the history of the region and the transformation from the Byzantine to the Islamic Empire. An essay in the journal of the Arab Language Academy, for example, hailed the career of Swiss scholar Max van Berchem (b. 1863) and his work on medieval Arabic inscriptions and Islamic archaeology. This mode of thinking had emerged in the late nineteenth century. An 1893 story published in *al-Hilal* on the Umayyad mosque in Damascus compared its structure to those of temples in Palmyra (Tadmor) and Baalbek and to Byzantine churches, in addition to drawing on information provided in Arab chronicles.

Nonetheless, the attitude toward Western archaeologists was ambivalent. On the one hand, as we have seen, there was much appreciation for Western archaeologists such as Breasted, who related to the Arabs and the Egyptians something of value about the history of their own forefathers and had managed to establish new facts based on their scientific inquiries. In this regard, it is worthwhile to note that the journals referred to Western scholars of the Middle East (ancient and Islamic), such as Howard Crosby, Charles James Lyall (1845–1920; noted for his studies of Arabic and Arabic poetry), and John P. Peters; quoted articles from the *Journal of the American Oriental Society*; and reported on conferences of Oriental languages. This tendency was even more pronounced in the journal of the Arab Language Academy, which informed its readers about works relating to Arabic studies and Middle Eastern history. It cited well-known scholars of the Middle East, such as Louis Massignon, Michelangelo Guidi, Carlo Alfonso Nallino, Martin Hartmann, Carl Brockelmann, David Samuel Margoliouth, and Richard Gottheil, and essays in *Revue du Monde Musulman*, *Journal Asiatique*, and others. When Ignác Goldziher (1850–1921) died, the journal published a long obituary that reviewed the scholar's life and his important works on Islam and on Judaism. The obituary opened with the following words:

> Semitic Studies and, moreover, Arabic in general and our academy in particular, grieves the loss of the great Hungarian scholar and Orientalist (*Majallat al-majma' al-'ilmi al-'arabi* 1 [1921]: 387).

On the other hand, during 1919–1921 suspicion toward archaeologists surfaced, growing out of the notion that objects which should have remained in Arab possession were being taken out of the region. This duality was evident as early as the late 1880s. Reports in *al-Muqtataf* from that time about Egyptian obelisks removed to Paris, London, and New York did not critique the decision to remove these objects. Yet another perspective had already taken root. In 1878 Jamil al-Mudawwar prefaced his series of articles on the history of Assyria and Babylon by lamenting the fact that entire cities could be built from the ancient Middle Eastern antiquities taken to Europe. "The treasures of our forefathers," he wrote, now decorate European cities. When he realized that the peoples of the region were unaware of the great value of these objects, and the magnificent history related to them, he decided to tell the story of Assyria and Babylon.

This sense of peril naturally increased following the occupation of the Middle East during World War I by the European powers. A story from 1920 on items transferred from Iraq to England in the section "scientific news" of *al-Muqtataf* opened with the following words: "Iraq is the land of the Babylonians and the Assyrians. Their mighty kingdoms left artifacts equal to those of ancient Egypt. As soon as the British occupied Iraq, their scientists began to excavate [the country's] antiques and had [already] sent to their country over 30 boxes." Similarly, in a story published in a 1920 issue of *al-Hilal* about the ruins of Aqar Quf, the author expressed his anxieties thus:

> Our hope is that the occupying government guards this old, magnificent, ancient site ... and that it invests the utmost effort to protect it ... and build big fences around it.... For if not ..., the country would be devoid of one of the most glorious things left to it from its forefathers, after it had survived the most difficult of times (*al-Hilal* 28 [1919–1920]: 95).

Finally, the attitude toward archaeology was connected to the rise of a new institution, the museum. Museums already existed in Cairo and Istanbul and were marked both as national sites preserving the cultural heritage of the region and its golden ages, and as institutions of great learning and science. A poem published in *al-Hilal* by Ahmad Muhammad Agubi, "The Egyptian Museum" [*al-Mathaf al-misri*], was written in the imperative. The speaker urged his imagined reader to go to the Egyptian Museum and explore the heritage of the "past fathers," to learn about the great king Ramesses II, to think of the ancient engineers who were able to build such spectacular structures, and to appreciate the days when Egypt was a great civilization, while nothing was known about Paris or London, a time when Egypt was a site of pilgrimage for the people of the West:

> Ask Pythagoras, ask Solon about the sciences they transferred from [ancient Egypt] to the West and about what they had written (*al-Hilal* 28 [1919–1920]: 439).

The reader was then ordered to contemplate what Egypt had given the world, in terms of science and culture. Those are our fathers, the speaker concluded, and we should follow their path. The visit to the museum, as represented in the poem, is not a passive experience during which the visitor merely observes objects. It is a learning experience in which differences between past and present collapse and the visitor (who is naturally an Egyptian) learns to rejoice in his homeland's splendid past. The visitor, moreover, begins to feel that his culture is the culture that shaped Western civilization.

THE WORLD OF THE SEMITES

In addition to archaeology and ancient Middle Eastern history, a key theme in the journals of which we have been speaking was the study of Semitic languages and the development of writing in the Middle East, seen as one of the region's most important contributions to global civilization. This theme was reiterated in almost every year of the publication of *al-Muqtataf* and *al-Hilal*. Essays took up the relationship between the various Semitic languages, providing information about Akkadian, Ugaritic, Phoenician, Aramaic, Hebrew, Syriac, Arabic South Arabian, and Ge'ez. Non-Semitic languages were considered as well. In discussions of the history of writing and methods of inscription, comparisons were commonly drawn between cuneiform, Egyptian hieroglyphs, and the Phoenician alphabet. *Al-Muqtataf*'s very first issue in 1876 included a lengthy article on the Himyarite language. In order to explain what this South Semitic language was, the editors included an introduction to the history of Semitic languages. The article was supplemented with photos comparing the Himyarite alphabet to the Arabic alphabet. In 1893 *al-Hilal* published an extensive essay on the history of writing that provided illustrations, tables comparing different Semitic alphabets, and photographs and another essay on the origins of language. These studies were also published in 1919–1921.

The writing about Semitic cultures and languages was related to the development of nationalist ideology. With the rise of Arab nationalism, languages began to be connected to ethnicities and the particular characteristics of the peoples who spoke them. The European division between Aryans and Semites was known to Middle Eastern intellectuals and they used it for their own purposes. The term "the Semitic Nation" [*al-umma al-samiyya*] was used in *al-Muqtataf* early in 1920 in an essay which reviewed the various migrations of Semitic peoples within the Middle East. For this reason, the determination of which languages belonged to the family of Semitic languages and which did not had important repercussions for writers. Combining national and cultural concerns, many articles cited the traditions about the children of Shem — Elam, Ashur, Aram, Arpachshad, and Lud — in identifying them as the progenitors of the nations of the regions. At the same time, writers were rarely satisfied with these traditions alone. It was crucial for writers, for example, to establish whether the inhabitants of Syria prior to the arrival of Alexander the Great were "pure" Semites and how the blood of these Semites was fused with "European blood" following the arrival of Alexander. The usage of the categories of blood and race indicate a pride in the Semitic heritage of the region and an attempt to racially connect the Arabs with the region's ancient peoples.

Within this context, it is worthwhile noting that the journal of the Arab Language Academy printed a speech by the famed French scholar of Islam, Louis Massignon (1855–1922), in which he critiqued the assumptions about the Semites made by the French philosopher Ernest Renan. Massignon argued that Semitic languages, unlike Aryan languages, were spiritual. However, he tried to convince his listeners that this spirituality was connected to the fact that Islam did not spread by the power of the sword alone, but also with the assistance of science and knowledge. He went on to contend that the East, rather than the West, was the first locale where a correlation between science and religion took place.

Not only nationalism, but philology, too, played a role in the writings about Semitic languages. Iskandar Isa Maluf clarified the key point of interest in *al-Muqtataf*: "the analysis of nouns and their comparison with nouns similar to them" — the juxtaposition between languages and between forms within each language should be considered as guides for historians. Maluf concluded that the careful historian, the archaeologist, and the linguist were thus the sources for uncovering the secrets of the past.

Interestingly, many of the articles acknowledged the fact that the region was typified by a multiplicity of languages, Semitic and non-Semitic; articles about the languages spoken in ancient Syria in particular stated this clearly. An article published in *al-Muqtataf* about the Jews of Iraq discussed the rise of Jewish Babylonian Aramaic in addition to the usage of Hebrew. Its author, the Iraqi intellectual Yusuf Ghanima, discussed how Babylon became a new center for Jewish national history [*tarikhuhum al-qawmi*] and he noted the changes that had occurred in the languages spoken by the Jews following their time in Babylon. In essays about the development of Arabic, language itself was seen as dynamic and changing from context to context and from time to time. Essays consequently reviewed the effects of Greek words on Arabic. They explained, for example, how fundamental Islamic terms, like *zakat* ("alms giving," one of the Five Pillars of Islam), originated from Greek terms. Two lengthy essays in the journal of the Arab Language Academy further noted the extent to which foreign words had been absorbed into Arabic, underlining the changes that Arabic went through following its interaction with the languages of the Persians, Byzantines, and the Copts as well as the speakers of other Semitic languages in the region. The connections between old and new languages were drawn in articles about the future of Arabic, which considered the modes in which the past could influence the present and the future. In the years 1919–1920 *al-Hilal* ran a survey addressed to various scholars (Arab, Turkish, European, and American) about the future of the Arabic language. Muhammad Kurd Ali called for the collection and publication of all that had been written by Islamic scholars on the Arabic language from the eighth century onward for readers to familiarize themselves with this scholarship.

The accounts about Semitic empires and languages were influenced by, modeled after, and ran parallel to stories and essays which attempted to convince Egyptians that the glorious pharaonic heritage was relevant to their identities. For example, a 1921 article about the illnesses of the ancient Egyptians by Hasan Kamal, which mostly summarized studies by M. A. Ruffer, suggested that "every country has sicknesses that spread in it according to its nature and its conditions of living, and this is the case in Egypt." The information in this essay, the author promised, would be valuable to every doctor and especially to the author's Egyptian brethren. Though pieces on ancient Egypt were directed at the modern Egyptian readership in particular, they were seen as relevant for all the peoples of the Middle East and for global history. In 1893, *al-Hilal* had run a story on the biography of Ramesses II as part of a long series about great men in history, from both the East and the West. Articles on the relationships between the Egyptians and the Phoenicians appeared in *al-Muqtataf* in 1921. Another essay published in the journal of the Arabic Language Academy on the histories of museums and scholarly gatherings tried to find the origins of such institutions not only in ancient Greece but also among the Egyptians, the Assyrians, and the Babylonians, as well as the Indians and the Chinese. This essay went on to discuss the literary markets and poetry gatherings in pre-Islamic Arabia and the literary and cultural salons in the palaces of the caliphs and the emirs. Cultural reciprocity thus influenced the cultures of the East and the West taken together, as well as those of particular kingdoms and states in the Middle East.

Writing about bilingualism and the changes in languages also reflected the sociocultural milieu of Arab writers. Most writers were bilingual or trilingual; in addition to Arabic, they were usually familiar with at least one

Western language, and many more knew Turkish. For the Christians, their liturgical language was often different from their spoken language. They lived in a world where Western education and knowledge of Western languages was an important vehicle for social mobility and where Arabic itself was changing and being reformed. Thus the exploration of the earliest stages of Arabic, as well as historical periods during which the region's peoples interacted with and influenced each other, appealed to contemporary readers. Moreover, the emphasis on reciprocity and cultural exchange in the ancient Middle East unsettled the narratives about pure blood, since these (often romanticized) accounts emphasized that integration of cultures was a key component of the region's history and culture.

CONCLUSIONS

This essay has been intended to provide a general sense of what educated Arabs knew about ancient Middle Eastern history and archaeology when Breasted visited the region in 1919–1920. It has dealt mainly with representations and romanticized notions of the past. In fact, some of the articles printed in *al-Hilal* and *al-Muqtataf* were not scientifically accurate or particularly profound. For example, the number of Middle Eastern peoples assumed to be "Semites" was quite inflated, as it included as many kingdoms and peoples as possible. The attempt to weave together the Bible, the Quran, and modern archaeology was not always successful. Of significance, however, is the fact that when Breasted visited the region, educated Arab elites were curious about their histories, languages, and pasts. This openness engendered an ambivalent approach to Western archaeologists, linguists, and historians. On the one hand, there was much appreciation of their work, and on the other, there was a fear that local antiquities would be looted and that Arabs knew far less about their histories than did foreigners. Arab intellectuals chose to selectively appropriate and hybridize bodies of knowledge which became available to them through the American universities in Cairo and Beirut, and their studies at European universities. Moreover, though the states in which these intellectuals were living (Egypt, the Ottoman Empire, and after 1921, the states in Iraq, Syria, Lebanon, Palestine, and Trans-Jordan) may not have displayed a keen interest in preservation and excavation, this did not discourage them from engaging in the discovery and the study of the ancient heritage of their countries. They were immersed in secularizing discourses regarding language, culture, and history. Furthermore, while criticisms of Zionism were already appearing in Arabic cultural magazines, other articles — which very respectfully described Jewish history, the Hebrew language and the Bible, and the historical connections between Jews and Arabs — were being published as well.

In recent years, problematic theories about "the clash of civilizations" have perpetuated certain cultural stereotypes about Arabs and Muslims as being hostile to Western modernity. Actions by fundamentalist regimes, such as the destruction of the Buddhas of Bamyan in 2001 by the Taliban, have served to solidify these stereotypes. Exploring the complex ways in which Arab intellectuals looked at their own history during the period of 1919–1921, their pluralistic approach to cultural affairs, and the multifaceted modes they used to draw connections between the East and the West might offer a certain remedy for such theories.

7. EPILOGUE: AN APPRAISAL OF THE FIRST EXPEDITION

EMILY TEETER

As Breasted stated in his proposal to Rockefeller (*Appendix A*), the aims of the inaugural expedition of 1919–1920 were to purchase antiquities for a new Oriental Institute and to select sites for future excavation. The following remarks explore how the expedition met, or did not meet, Breasted's goals in laying the groundwork for the Oriental Institute up until the time of his death in 1935.

Once back in Chicago, Breasted delivered the Report of the First Expedition of the Oriental Institute of the University of Chicago to University President Harry Judson (*Appendix B* [Report]). He divided the Report into a summary of the trip itself including purchasing activities and "Relations with Governments," followed by the "Political Mission to England," recounting the recommendations that he gave to Lord Curzon of the British Foreign Office on his way back to the United States. This was followed by "Opportunities and Recommendations" that covered desirable objects to acquire for the museum and his recommendations of sites for future excavation.

He reminded the president of the value of the 1919–1920 expedition and the "obligation" and opportunities that it opened to the University. He reported that the trip had been invaluable for developing an overall strategy for the future, sometimes in the most practical ways:

> The facts regarding prices of labor, the season when labor is free to leave flocks and fields, the possibilities for disposing of excavated rubbish, and all items of information essential to carrying on excavations at all important points in Mesopotamia, Syria and Palestine were carefully collected (J. Breasted, Report, p. 19).

He also cited the important connections he had made with "controlling authorities" and with "sheiks and natives of influence" (J. Breasted, Report, p. 19).

Breasted's vision for the future of the Oriental Institute was so expansive that it virtually ignored the future efforts of other archaeologists from America or any other nation, making his plan for the recovery of the history and cultures of the ancient Middle East primarily a Chicago project:

> Before the whole recoverable story drawn out of every available mound is in our hands, it may be indeed a century or two.... I am confident that with sufficient funds and adequate personnel <u>it will be possible in the next twenty-five or thirty years, or let us say within a generation, to clear up the leading ancient cities of Western Asia and to recover and preserve for future study the vast body of human records which they contain</u> (J. Breasted, Report, p. 27).

Breasted rationalized his vision as an opportunity for America:

> ... I cannot but see in the recovery and study of this incomparable body of evidence America's greatest opportunity in humanistic research and discovery.... I can only add a reference to the impoverishment of European governments and their entire lack of men to do this work.... This complete paralysis of Europe in oriental research thus not only shifts a grave responsibility upon the shoulders of America, but at the same time enlarges our own opportunity as never before (J. Breasted, Report, p. 27).

A major recommendation of his Report (p. 23) was an aggressive campaign of excavation. He described the situation, typically equating the study of history with the hard sciences:

> The Near East is a vast treasury of perishing human records, the recovery and study of which demand a comprehensive plan of attack as well organized and developed as the investigation of the skies by our impressive group of observatories, or of disease by our numerous laboratories of biology and medicine.... the ancient city itself with its streets, buildings, walls, gates, water-works,

drains and sanitary arrangements is a fascinating and instructive record of human progress and achievement, which must be studied, surveyed and recorded, in the same way the geology, botany and zoology of the Near East must be studied to reveal the character of the habitat and resources of the earliest civilized communities of men.

MUSEUM PURCHASES

The acquisitions that Breasted made during the expedition were significant for the growth and scope of the collections of the Oriental Institute and the Art Institute of Chicago. Comparing the difference in quality of the purchases from this trip to those from his first visit to Egypt in 1894 is astounding. Breasted showed far greater confidence in his selections as well as a talent for negotiating with dealers. Although he never considered himself a connoisseur of Egyptian art, he developed a keen eye for objects of beauty that were also highly instructive.

Breasted's enthusiasm for continuing to acquire objects for the museum is reflected in the ambitious recommendations presented to President Judson in his Report. The first was Lord Wimborne's collection of Assyrian reliefs installed as the "Nineveh Porch" at his estate in Canford, Great Britain (for these reliefs, see Russell, *From Nineveh to New York*). The collection was ultimately acquired (with funding by Rockefeller) by the Metropolitan Museum of Art in New York. However, in retrospect, Breasted was not disappointed, for as the dealer Kelekian (who offered the collection) wrote, "I proposed them to Prof. Breasted of the Chicago University, and his answer was that for the price I am asking, he could make excavations in Assyria for ten years."[10]

His second recommendation was the purchase of the George Ford Collection of materials excavated at Sidon, including important Greco-Phoenician sarcophagi. That collection ultimately was purchased by the National Museum in Beirut. His third recommendation, the acquisition of the Egyptian Coptos decrees was similarly unsuccessful.

EXCAVATIONS

Breasted's recommendations for excavations were grand, but overall they were far more fruitful than his suggestions for acquisitions. It is not surprising, considering that he was an Egyptologist, that the first excavation of the Oriental Institute was in Egypt.

In the Report (p. 23), he recommended that the Institute excavate Memphis because it was a chronologically comprehensive site that reflected "the entire range of ancient civilization." He continued (pp. 23–24):

> This vast cemetery has thus far only been nibbled at. The Egyptian Government, while reserving it for excavation by its own Department of Antiquities, has in vain endeavored to meet the obligation thus assumed. By actual computation by one of its own staff, it will take the Egyptian Government five hundred years to complete the excavation of the Memphite cemetery at the present rate of progress. Conferences with the Milner Commission gave me an opportunity to put this situation clearly before them and they concluded that excavation by the Egyptian Government, if not discontinued should at all events be discountenanced in favor of a policy of yielding to private initiative the chief responsibility for rescuing such enormous bodies of records for scientific use.

However, the first excavations were not carried out at Memphis, but rather in western Thebes at Medinet Habu. Just two years after his return, he appointed Harold Nelson of the American University of Beirut as Director of the new Epigraphic Survey to copy and publish the reliefs and inscriptions on the great temple of Ramesses III (fig. 7.1). Today, the Epigraphic Survey continues to thrive in Luxor from its headquarters at Chicago House. In 1926, the Epigraphic Survey was joined by the Architectural Survey (1926–1933) whose mission was to clear the temple so that the epigraphers could access the wall decoration and to study the architecture of the structure (fig. 7.2).

---

[10] Russell, *Nineveh to New York*, p. 135. The collection was offered to Breasted for $450,000 (about $5.7 million in today's dollars).

FIGURE 7.1 Members of the Epigraphic Survey copying inscriptions on the temple of Ramesses III in western Thebes, 1927 (OIM photograph P. 14834)

FIGURE 7.2 Breasted, Harold Nelson, and Uvo Hölscher, Director of the Achitectural Survey, watch the sarcophagus of Harsiese being raised from his tomb at Medinet Habu, 1929 (OIM photograph P. 18763)

Some of the Oriental Institute projects in Egypt were based on Breasted's trip, but others were implemented by further funding by John D. Rockefeller Jr. In 1929 Breasted accompanied Rockefeller on a trip through the Nile Valley piquing his interest with specific bodies of material in Saqqara and Abydos. Results were the Saqqara (Memphis) Expedition 1930–1931 under the direction of Prentice Duell, a professor in the Art Department of Bryn Mawr who had worked on the recording of Etruscan tombs. Typically, the scientific staff was housed in a beautifully landscaped dig house just outside the archeological zone. Their initial and ambitious plan was to publish the mastaba tombs of Ti, Ptahhotep, Mereruka, Kagemni, Idut, "and one or two smaller mastabas." Ultimately, only half of the mastaba of Mereruka was published in two oversize folio volumes, and the Saqqara Expedition was disbanded in 1936. The publication of the reliefs and inscriptions from the temple of Seti I at Abydos (fig. 7.3) was also due to Rockefeller. When he visited the site he was impressed with the beauty of the wall scenes, and he expressed an interest in seeing that the material was published in "the most magnificent form." The resulting four folio volumes were a joint publication of the Oriental Institute and the Egypt Exploration Society (fig. 7.4). Breasted was also responsible for Rockefeller providing the funding for the publication of the exquisite facsimile copies of Theban tomb scenes made by Nina and Norman de Garis Davies. Out of this project came the three volumes entitled *Ancient Egyptian Paintings* that appeared in 1936 (fig. 7.5).

Another direct result of the trip was the excavation of Megiddo (fig. 7.6). In his Report (p. 26), Breasted wrote that "the British authorities have assured me that they will reserve this place for excavation by the University of Chicago," and that "Prof. Garstang, Director of the British School in Jerusalem voluntarily offered to hold in reserve for the University of Chicago, the splendid fortress city of Megiddo" (p. 20). Ironically, Breasted had not visited this site because "a stupid guide misled us so that we failed to reach Megiddo itself..." (p. 17). Excavations were started in June 1925 under the direction of Clarence Fisher. Although Breasted declared that the goal was "not the discovery of museum pieces but rather an exhaustive salvaging of the available evidence" (J. Breasted, *Oriental Institute*, p. 240), the excavation uncovered a wide range of objects that documented occupation of the site from 3500 to 332 BC. Some of the objects awarded to Chicago in the division, especially the Megiddo ivories, are today among the highlights of the collection of the Oriental Institute. Breasted's interest in aerial survey was furthered at Megiddo when the team fitted a meteorological balloon with a camera and sent it aloft over the site (fig. 7.7).

Work in Iraq figured highly in Breasted's recommendations. In the Report (p. 19) he recommended that the University dig at Nimrud, reporting, "the Civil Commissioner at Baghdad ... assured me he would welcome an expedition of the University of Chicago, which might desire to excavate in Mesopotamia, and that we could count upon having the site of Nimrud ... if we desired it." He optimistically predicted (p. 25) "A programme including the excavation of the gates of Khorsabad and the whole of Nimrud, could be carried out in a few seasons."

He noted that Nineveh was held by the British, although "they have thus far met by nothing more than a little haphazard grubbing." Breasted (Report, p. 25) urged the University to assume control of the site through collaboration with the British:

Although they are still reserving Nineveh for themselves, the British are already anxious to

FIGURE 7.3 Epigraphers at work in the tomb of Mereruka at Saqqara, ca. 1934 (OIM photograph P. 24466)

FIGURE 7.4 Sety I offers a pectoral and collar to Osiris. Calverley and Gardiner, *The Temple of King Sethos I atAbydos* (vol. 1, pl. 16, 1933). A joint publication of the Egypt Exploration Society and the University of Chicago

FIGURE 7.5 Scribe registering Nubian tribute. Davies and Gardiner, *Ancient Egyptian Paintings* (vol. 1, pl. 16, 1936). Published with the financial support of John D. Rockefeller Jr.

FIGURE 7.6 The excavation of Megiddo, 1931 (OIM Megiddo field negative A460)

FIGURE 7.7 Aerial photography at Megiddo was made using a meteorological balloon, 1931 (OIM photograph P. 18637)

make some form of combination with Americans in order to eke out their own meagre resources. A few successful seasons by an American expedition at the gates of Khorsabad and the palaces of Nimrud, would enable us to put the British in a situation where something would have to be done by them to make the records buried in the great Assyrian capital accessible to science. Diplomatic handing of this situation, would in my judgment put our expedition in command at Nineveh, where a decade of successful work would enable us to restore to modern knowledge the vast treasury of human records and human handiwork now lying buried in the greatest imperial capital of Western Asia. No task more pressing nor more illustrious in its achievement than the excavation and recovery of this magnificent Rome of Western Asia is to be found in the whole range of humanistic research.

Nineveh and Nimrud were not to be. The first site selected for excavation was Khorsabad, a site that had impressed Breasted highly in 1920. It was excavated for seven years, from 1928 to 1935, initially under the directorship of Edward Chiera, who was succeeded by Henri Frankfort when Chiera was recalled to Chicago to head another of Breasted's dreams — the great Chicago Assyrian Dictionary project (see below). In 1931, another team was sent to the Diyala region east of Baghdad to work at Khafaje, Tell Asmar, and Tell Agrab (fig. 7.8). The Iraq Expedition was staffed with a stellar group of academics including Pierre Delougaz, Henri Frankfort, Thorkild Jacobsen, and Seton Lloyd. In addition, these excavations of Assyrian and Sumerian sites brought a wealth of material to the Oriental Institute, foremost being reliefs and monumental sculpture from the palace of Sargon II at Khorsabad (fig. 7.9) and a selection of Early Dynastic worshipper figurines from Tell Asmar (fig. 7.10).

In his Report, Breasted advocated excavations at Kadesh, a site in Syria that he knew well from his 1920 visit and from his 1903 publication of the great battle between Egypt and Mitanni. He argued that Kadesh, as

FIGURE 7.8 Members of the Iraq Expedition on the steps of the dig house at Tell Asmar in the Diyala, February 24, 1931. Left to right: Pierre Delougaz, Thorkild Jacobsen, Rigmore Jacobsen, Mary Chubb, Rachel Levi, Henri Frankfort, and Conrad Preusser (Tell Asmar field negative A213)

FIGURE 7.9 Excavation of a gateway at the palace of Sargon II, Khorsabad, 1934 (OIM photograph P. 30952)

a "composite civilization made up of Babylonian, Hittite and Egyptian elements, which developed in Syria and found its way to Europe through Asia Minor by land, and the ports of Phoenicia by sea" (p. 25) was an important part of the overall strategy of the Oriental Institute to trace the westward movement of Babylonian civilization toward Europe — part of his core goal of demonstrating how the roots of Western civilization lay in the ancient Middle East. Kadesh was ultimately not excavated by the Oriental Institute, but much later by a British team. Similar research objectives under the direction of Calvin McEwan would be realized by the Amuq projects in the plain of Antioch, an area too dangerous for Breasted to visit in 1920.

Although Breasted did not visit Turkey in 1920, by 1926 the Institute sent out the Anatolian-Hittite Expedition under the direction of H. H. von der Osten. The team surveyed central and southeast Turkey, where they discovered "several hundred ancient sites" (J. Breasted, *Oriental Institute*, p. 277) (fig. 7.11). The expedition concluded in 1932.

Breasted also recommended two sites in Lebanon for excavation, but these plans never materialized. The first was Byblos on the coast of Lebanon. He favored the site because operations on the Mediterranean coast would be the final step in documenting the transmission of the cultures of the ancient Middle East to

FIGURE 7.10 Discovery of a cache of Sumerian worshipper statues at Tell Asmar, January 1934 (OIM photograph P. 23290)

FIGURE 7.11 The staff of the Hittite Expedition that surveyed sections of central and southeast Turkey by car, discovering hundreds of ancient sites. Imatlü, 1926 (OIM photograph P. 13299)

Europe. This area, like Iraq, according to Breasted (Report, p. 24), was just waiting for scholars of the Oriental Institute: "... we visited [Byblos] and found it laying ready for excavation." He envisioned this as a joint project with excavations at Memphis because of the ancient ties between the two sites: "This illustrates the necessity of a group of expeditions and the correlation of their results." He also suggested excavations at Sidon in Lebanon. He had seen the important collection of Greco-Phoenician antiquities in the collection of Dr. Ford of the American Mission (which, as already mentioned, he wanted the University to buy). But both Breasted and Ford thought Sidon would still be productive, as Breasted (Report, p. 24) wrote: "[Dr. Ford] assured me he would be very glad to have us complete the clearance of the tombs on his property."

The Oriental Institute continued to develop Breasted's initial vision with excavations across the Middle East in the 1930s. The Institute undertook excavations in southeast Turkey (then part of Syria) in the Amuq Valley (fig. 7.12) with the Amuq Expedition (also called the Syrian-Hittite Expedition) (1931–1938) under Field Director Calvin McEwan. Large-scale operations were opened at the Persian capital city of Persepolis in 1931 (fig. 7.13). The permit to excavate this highly sought after site was granted on the basis of the growing reputation of the Oriental Institute which, as a result, became the first foreign mission to be granted an excavation permit after the French monopoly to work in Iran was revoked by Reza Shah in 1929.[11] Work commenced under the direction of Ernest

FIGURE 7.12 Statue bases in the form of snarling lions discovered by the Amuq Expedition at Tell Tayinat in southeast Turkey, 1934–1935 (OIM photograph P. 27120)

[11] See J. Breasted, *Oriental Institute*, p. 311, where he states that Herzfeld worked at Kuh-i-Khwadja (Kuh-e Khajeh) and Pasargadae on behalf of the Forschungsgemeinshaft der deutschen Wissenschaft. This is only partially correct. His work at Pasargadae was on behalf of the German institution, but the permission to work at Kuh-i-Khwadja was granted directly to Herzfeld — not through any European organization. I thank Abbas Alizadeh for this information.

FIGURE 7.13 Excavations on the terrace of the Apadana at Persepolis, ca. 1933 (OIM photograph P. 22257)

Herzfeld, who in 1935 was succeeded by Erich Schmidt, who led the work until its conclusion in 1939. Divisions from the excavation include monumental sculpture, and highly important groups of administrative tablets.

The pace of Chicago-sponsored archaeology was frenetic. In 1931, there were eight excavations and two epigraphic projects simultaneously underway.

OVERSEAS HEADQUARTERS

While in Cairo, Breasted visited the French Institute. Taking that as his inspiration, he mused what he might be able to do along the same lines. By the time he wrote his Report to Judson, an ambitious plan had crystallized. He recommended that the data from the excavations

> would therefore have to be gathered together at a common center where the process of study, correlation and publication could be steadily carried on. For this purpose there would eventually be necessary a winter headquarters at Cairo and a summer headquarters on the high cool slopes of Lebanon overlooking Beyrut. These two centers would together form an ORIENTAL INSTITUTE HEADQUARTERS on the ground and together constitute a common center furnishing both <u>administrative</u> and <u>investigative</u> direction of the work throughout the Near East (Report, p. 26; emphasis in orignal).

These centers would provide working quarters for a centralized group of "investigators" who would "receive, classify, correlate, study and publish the facts and sources discovered in the field...." Perhaps his lack of excavation experience led him to assume that the field directors of projects would not object to having this proposed group in Cairo and Beirut analyze their finds, leaving them as mere excavation machines.

Rather than establishing two centralized research centers, the Oriental Institute built a series of residence-work centers at each of their major excavations. These centers offered comfortable, even lavish accommodations by modern standards, each staffed with locals to cook, clean, and attend to the needs of the researchers (figs. 7.14–17). Extensive libraries in Luxor and Megiddo enabled the scholars to interpret their finds in the field (fig. 7.17) and to write their reports.

Breasted's vision was manifested in Chicago as well. In 1921, the Chicago Assyrian Dictionary was founded. Today, the project is nearing completion with its twenty-first volume. The American headquarters of the Oriental Institute were moved into a large and newly designed building in 1931 (figs. 7.18–19) that serves as a reminder of the influence of James Henry Breasted and his bold plan to document the history of mankind.

FIGURE 7.14 The dig house at Megiddo, 1934 (Megiddo field negative A1181)

FIGURE 7.15 The headquarters for the Diyala excavations at Tell Asmar, 1932 (Tell Asmar field negative A417)

FIGURE 7.16 "New" Chicago House, built in 1930 on the east bank of the Nile. The complex was designed by the same architectural firm that planned the Chicago headquarters (see fig. 7.18) (OIM photograph P. 22323)

FIGURE 7.17 The library at Chicago House, ca. 1933. The library continues to be
a center of academic activity in Upper Egypt (OIM photograph P. 18875)

FIGURE 7.18 The Chicago headquarters of the Oriental Institute, opened in 1931. The building contains laboratories, offices, a reference library, museum preparation areas, storage, 16,000 square feet of museum galleries, an auditorium, and two small classrooms (OIM photograph P. 18730)

FIGURE 7.19 View of the Egyptian gallery at the Oriental Institute, 1937. The Assyrian *lamassu* (winged bull) from the excavations at Khorsabad can be seen at the end of the hall (OIM photograph P. 29038)

APPENDIX A

PROPOSAL FOR THE FOUNDING OF THE ORIENTAL INSTITUTE
SUBMITTED BY JAMES HENRY BREASTED
TO JOHN D. ROCKEFELLER JR., FEBRUARY 1919[1]

[Page -1- of original]

PLAN FOR THE ORGANIZATION OF

AN ORIENTAL INSTITUTE AT

THE UNIVERSITY OF CHICAGO

1. THE OPPORTUNITY AND THE OBLIGATION.

Within a few weeks the ancient lands of Western Asia where civilization and the great world religions were born, have for the first time in history been rendered safe and accessible to research and investigation. Here lie the unexplored areas of history. The study of these lands is the birthright and the sacred legacy of all civilised peoples. Their delivery from the Turk brings to us an opportunity such as the world has never seen before and will never see again. Our Allies in Europe are financially too exhausted to take advantage of the great opportunity. This makes the opportunity and the obligation all the greater for us in America.

2. THE IMMINENT DESTRUCTION OF THE MONUMENTS AND THEIR INACCESSIBILITY EVEN IN THE MUSEUMS.

It is evident that the opening of Asia Minor, Syria, Palestine, Mesopotamia and Babylonia to modern business and to enlightened exploitation in mining, railroad-building, manufactures and agriculture, means the rapid destruction of the great ruined cities and buried records of early man, with which these lands are filled. They must be studied soon before they are lost forever.

Great numbers of accessible monuments in the Near East are still unpublished, and the museums of Europe are likewise great storehouses of unpublished documents. Every season in western Asia and Egypt a large body of new documents is turned up: some in scientific excavations, some in illicit native diggings, some by accident. In the hands of natives, documents of priceless value frequently knock about for months or years, and then perish. This happens all the time. In many cases the camera of the visiting archaeologist might have rescued the document in a few minutes, even if he was unable to buy it; or an hour's work would have produced a copy of it in his note-book.

[1] The original document is typed and resides in the Oriental Institute Archives. This transcription aims to reproduce the basic layout of the report without maintaining line breaks or pagination. Page numbers of the original document are given in brackets.

When taken out by alleged scientific excavators, the documents are often never published. In the last twelve years probably two or three thousand packing boxes have wandered from Egypt to the Museum of Turin, Italy, where their contents have been unpacked and installed. No account of these monuments or of the excavations which produced them has ever appeared or is likely to appear. Of such unpublished records there is therefore a vast and ever growing body.

Besides these, there are the unvisited sites of ancient cities, where much may often be saved by a mere examination of the surface. At the Hittite capital of Khatti in Asia Minor, Winckler on his first visit to the place kicked out with his boot-heel documents from the royal archives of the Hittite foreign office, which were lying only a few inches below the surface. Wagon loads of royal records lay just below. The result was the discovery of the materials which have made possible the decipherment of the lost Hittite language. At Sidon Dr. George Ford now has lying in his house ten or more splendid Phoenician sarcophagi of stone, discovered by accident. They have been lying there for years and no one has given them any attention. Many more such examples might be adduced.

3. THE NECESSITY OF COLLECTING THE AVAILABLE DOCUMENTS IN SYSTEMATIC ARCHIVES.

It is needless to point out that the university teacher is as unable single-handed to cope with a situation like this as would be the astronomer to study the skies without his observatory or his staff. In France the national Academy or the Government partially meets the difficulty by occasionally granting French savants a subvention for a visit to the Orient. This practice does not solve the difficulty, because the French savant's tenure of the subvention is temporary, and there is no common file or body of archives where the documents brought back are collected. They are thus scattered through the papers of a large number of different scholars who may at different times have held the subvention. The scattered fragments of man's story have never been brought together by any one. Yet they must be brought together by some efficient organisation and collected under one roof before the historian of today can piece together and reveal to modern man the story of his own career. A laboratory

[-2-]

containing all the available early human records in systematically arranged archives is as necessary to a study of man's career as an astronomical observatory with its files of observations and computations in the study of the career of the universe. It is evident that the methods and the equipment of natural science should be applied to the study of man, and the vast body of documents left behind must be as systematically gathered, filed and employed as are the observations of the astronomer.

4. THE NECESSITY OF REGULAR AND PERIODIC ACCESS TO THE AVAILABLE DOCUMENTS.

The astronomer is sometimes required to visit distant regions in order to make his observations. This is <u>constantly</u> true of the orientalist and ancient historian. To secure his materials, he must be granted the time and the funds to become a kind of permanent archaeological ambassador-at-large to the Near Orient. In this way the records resulting from a collecting activity covering many years might be brought together at one place. For he would then be able to visit at short intervals all the great sources of materials, whether in museums of Europe or in the Orient, and even carry on occasional explorations for new and perishing ruins and bodies of documents, like those found by Winckler at the Hittite capital. By means of the camera he could rescue forever large numbers of written documents and monumental ruins still unsurveyed. His photographs, journals, note-books, drawings and surveys, especially if he had an assistant to aid him in the field, would rapidly grow into a comprehensive group of documents. They would form a methodically collected body of historical archives, which he would spend all his available time in America in studying. Such a treasury of ancient records would soon become a focus, a clearing house for the correlation of all the prehistoric life and the various early civilisations grouped around the eastern end of the Mediterranean and thence at least as far as Persia. The final result would be a systematically built up documentary basis, such as exists nowhere else, for recovering the lost chapters of the career of man.

5. THE ORIENTAL INSTITUTE.

Housed in the Haskell Oriental Museum of the University of Chicago this great body of documents would constitute a historical laboratory which might be called THE ORIENTAL INSTITUTE. Just as the astronomical observatory requires a staff of assistants for the care of its records and files, so the Oriental Institute would need a small staff of helpers to keep the files in order, to arrange, accession and catalogue the various materials and documents. This would enable the director to maintain a constant general control of the whole great body of sources in the files of the Institute, and to devote his time in America to purely historical study of them.

6. THE AIM A GREAT HISTORY OF THE ORIGIN AND DEVELOPMENT OF CIVILIZATION.

The ultimate aim of all this organization would be to put together the story of man from the remotest ages, in order thus to trace as fully as possible his rise from Stone Age barbarism, through successive stages of advance, the emergence of civilization, the history of the earliest great civilized states, and the transmission to Europe of the civilisation which we have since inherited. In short the ultimate aim of such work must be the production of a great history of the Origin and Development of Civilization.

7. ADDITIONAL ADVANTAGES TO AMERICAN INSTITUTIONS AND SCHOLARS.

Frequent visits to the Orient, now so completely accessible, would give to the director of the Institute unusually favorable opportunities for purchasing original ancient documents and monuments at relatively low prices. For this outlay he could without doubt secure funds outside of the budget of the Oriental Institute from other sources. Indeed there are constantly traveling in the Near Orient in normal times Americans of means, who are quite ready to assist institutions in talking advantage of urgent opportunities to acquire ancient monuments and documents in the hands of oriental dealers. Such purchases would build up the Oriental Museum of the University, and at the same time furnish materials for original investigations by all members of the Department of Oriental Languages. Thus the materials from the field would be brought into the study rooms of all the other members of the Department also, and furnished with new documents in this way, the whole Department would be transformed into a much more productive scientific body.

[-3-]

The presence of an American representative who is frequently on the ground in many different regions of the Near Orient would be of substantial value in other ways,- especially because it would keep Americans constantly informed of favorable opportunities, not only for purchases as just indicated, but also for newly discovered openings or sites for excavation, to which the attention of American museums or interested patrons could be called. Eventually reports of the Oriental Institute on present-day conditions in the Near East might also be of value to our government, to our educational and relief organizations, and even to our business men.

8. THE ORIENTAL INSTITUTE NOT AN EXCAVATION ORGANIZATION AND HENCE A MODEST BUDGET.

While the Oriental Institute might accomplish much in suggesting and encouraging excavation, its plan does not contemplate supporting from its own budget any costly excavation campaigns. Its budget is therefore a modest one. It could be set going for about $10,000 a year. An endowment of $250,000 would therefore be necessary to launch it. As its collections of documents increased and it grew into larger opportunities of work, as it undoubtedly would, it would soon need a larger income. If the General Education Board would grant the University of Chicago an endowment of $250,000 to set the Oriental Institute going, it would without doubt be possible, as the need arose, to raise additional endowment for it from other sources.

9. MINIMUM BUDGET FOR PRELIMINARY AND EARLY STAGES.

Salary of Director (supplementing University salary) $1,500.00

Salary of Curator in charge of all files 2,000.00

Salary of stenographer & cataloguer (one person at first) . . 1,500.00

Draughting, photographing and supplies for same 1,000.00

Cases and files as records grow 500.00

Bulletins, reports and office supplies 1,000.00

Traveling expenses 2,500.00

Total. . . . $10,000.00

The above budget would suffice to inaugurate the enterprise. The organization would be much more effective if it could include a permanent field assistant, residing probably in Aleppo, or perhaps Cairo, with two rooms for headquarters, used chiefly for storage of field equipment. As the bulk of the files and records increased, additional help would be needed to care for them, and a staff photographer to furnish prints and care for the great file of negatives, precisely as is done in an astronomical observatory. For this expansion I am confident I can secure the endowment needed, once the enterprise has demonstrated its efficiency and its possibilities. In order to make this demonstration the General Education Board is asked to contribute the above initial endowment.

Respectfully submitted,
(Signed) James H. Breasted

APPENDIX B

REPORT ON THE FIRST EXPEDITION OF THE ORIENTAL INSTITUTE OF THE UNIVERSITY OF CHICAGO[1]

[Page -1- of original]

REPORT ON THE FIRST EXPEDITION OF THE ORIENTAL INSTITUTE
OF THE UNIVERSITY OF CHICAGO

President Harry Pratt Judson,
  The University of Chicago,
    Chicago, Illinois.

Sir:-

I have the honor to present herewith the director's report on the first expedition of the Oriental Institute of the University of Chicago.

The report is divided into the following sections and subsections:

I. EUROPE

1. ENGLAND
2. FRANCE

II. EGYPT

1. SCIENTIFIC WORK
2. PURCHASES OF ANTIQUITIES
3. RELATIONS WITH GOVERNMENTS & PREPARATIONS FOR ASIATIC TRIP

III. ASIA

1. BABYLONIA
2. ASSYRIA
3. OVERLAND RETURN JOURNEY ACROSS THE NEW ARAB STATE
4. SYRIA
5. PALESTINE
6. PURCHASES OF ANTIQUITIES

---

[1] The original document is typed and resides in the Oriental Institute Archives. This transcription aims to reproduce the basic layout of the report without maintaining line breaks or pagination. Idiosyncratic spellings are maintained. Page numbers of the original document are given in brackets.

IV. POLITICAL MISSION TO ENGLAND

1. CAIRO EVENTS
2. CAIRO TO ENGLAND
3. REPORT TO THE BRITISH MINISTERS IN LONDON

V. SECOND TRIP TO PARIS

1. NEW PURCHASES OF BABYLONIAN AND ASSYRIAN RECORDS
2. LORD WIMBORNE'S ASSYRIAN SCULPTURES

VI. OPPORTUNITIES & RECOMMENDATIONS

1. PURCHASES RECOMMENDED
2. OPPORTUNITIES FOR EXCAVATION AND RESEARCH

[-2-]

I. EUROPE

1. ENGLAND

London Meeting of the Orientalists of England, France and America

The director sailed from New York on August 21st, 1919, for England and arrived in London on August 29th with the special purpose of participating as official representative of the American Oriental Society in the joint sessions of the Royal Asiatic Society, the Societe Asiatique, and the American Oriental Society, meeting in an international conference suggested by the eminent French Orientalist Emile Senart, President of the Societe Asiatique.

This meeting, which occupied a large part of the first week of September, proved a very profitable opportunity for meeting the leading orientalists of England and France and discussing with them comprehensive plans for coöperation in many ways, especially in excavation and exploration in the newly opened regions of the Orient. It was decided that this joint conference should be perpetuated as an annual event.

Collections Visited and Studied

An invitation to Highclere Castle in Hampshire furnished the privilege of examining the remarkable recent collection of Egyptian art made by the Earl of Carnarvon. The Carnarvon Collection is unrivaled in variety and sumptuous splendor. Besides this collection, those at the Ashmolean Museum of Oxford, University College in London, and the British Museum were studied. These studies were much interrupted by

Preparations for the Oriental Expedition

122

Collecting the material equipment, and making of special trunks for apparatus consumed a great deal of time. It was also exceedingly difficult under post-war conditions to ensure transportation for the equipment and personnel of the expedition to the Near East. Many days were spent on this problem alone. Letters from the British Leaders were of great assistance. These included introductions from Lord Allenby, the Earl of Carnarvon, Lord Porchester, Dr. Hogarth of Oxford, Sir William Garstin (Former Head of Egyptian Public Works), Dr. Hall of the British Museum. Eventually Sir Frederick Kenyon, Head of the British Museum, also cooperated very cordially. But in spite of the most cordial reception at the hands of British officials everywhere, it was still uncertain how far we might be able to penetrate Asia when the director left London for Egypt, by way of France and Italy on October 9th.

[-3-]

2. FRANCE

Collections Visited and Studied

Besides the Louvre and the Musee Guimet, the director visited the magnificent prehistoric collections in the National Museum at St. Germain. The distinguished director of these collections, M. Saloman Reinach was very cordial, and consented to extend his hearty support to a request of the University of Chicago for a permanent loan collection of the leading types of prehistoric stone weapons and tools from the magazines of the St. Germain Museum. Such a collection has already been loaned to the Ashmolean Museum at Oxford, and can be secured by the University of Chicago if the loan is requested by the President of the University in a letter addressed to the French Government (Ministere de l'Instruction Publique), and forwarded through the French Ambassador in Washington.

Purchases of Antiquities

Several days were spent going through the large collections of the Armenians, Kalebdjian Freres, and a selection filling some gaps in Haskell Museum collections was made and purchased. The most notable among these purchases was a papyrus copy of the Book of the Dead, written in hieratic and with numerous colored vignettes, the roll being about thirty feet long and twenty inches high.

III [sic]. EGYPT

Leaving Paris on October 17th and embarking from Venice, the director arrived in Cairo on October 30th after a journey of almost insurmountable difficulties at a number of points.

1. SCIENTIFIC WORK

The Cairo Museum

Attention was devoted chiefly to new accessions and discoveries still needing study. Among these the most notable monument was the fragments of the Royal Annals formerly known as the Palermo Stone. Although these new Cairo fragments had been twice, published first by Gauthier and then by Daressy, it was still possible to secure numerous new readings, especially a group of ten pre-dynastic kings of united Egypt, that is a pre-Dynastic dynasty, proving therefore that there was a long-enduring union of Egypt before the dynasties. These discoveries will necessitate a new publication of the famous document, the oldest known royal annals in human history, and the materials for this new publication have been brought home. The magnificent coffin of Ikhnaton, encrusted with gold and precious stones was also studied in company with Mr. Bull of the expedition, and the publication of the inscriptions corrected.

[-4-]

Excavations Visited.

Since the director's last visit at Gizeh very important new excavations had been made there. These were repeatedly visited and studied, for they furnish the earliest chapters in the history of architecture in stone. Dr. George A. Reisner's summer camp here was also visited and his remarkable discoveries of Egyptian jewelry from Napata, made for the kings of Ethiopia in the days of the prophet Isaiah, were inspected. Dr. Reisner has promised to contribute from his former excavations of the earliest cemeteries a prehistoric body for our Haskell collections.

Important new facts in architecture have been discovered by the Philadelphia Expedition at Memphis, where Mr. Clarence S. Fisher has uncovered a palace of Pharaoh Merneptah who lived toward 1200 B.C., the most probable date of the Hebrew Exodus. The palace had been destroyed by fire and Fisher found the great doors of the throne room burned to ashes and their heavy metal pivot hinges far out in the hall where they had dropped from the massive cedar woodwork as the blazing doors toppled over far out into the hall and carried the pivots along. It is rather interesting to recall that if Merneptah was really the Pharaoh of the Hebrew Exodus, this is the room where the Hebrew tradition would have placed the famous scenes between the Pharaoh and Moses and Aaron.

The extraordinary unfinished Fourth Dynasty pyramid at Abu Roash, where a colossal structural causeway still survives was visited in company with Lord and Lady Allenby, who asked me to ride out there on horseback. This ride, while it also furnished some archaeological observations, served also as an opportunity of presenting to Lord Allenby the needs of our coming expedition into Asia.

Similarly an invitation of Sir Robert Greg, Director-General of the Egyptian Foreign Office, to visit the excavations at Abydos and Tell el-Amarna, was a very valuable opportunity to bring our plans for further explorations in Asia before the British authorities. At the same time the official protection which Mr. Greg controlled made possible a visit in the dangerous region of Amarna, which would otherwise have been entirely out of the question.

On returning to Cairo the day before Christmas, the director found Mr. Ludlow S. Bull, Fellow of the Department of Oriental Languages, just arriving from America, the first additional member of the expedition to join the director in the Orient. Mr. Bull then took up studies in the Museum under my direction and accompanied me also in the inspection of excavations at Sakkara, Abusir, and Abu Ghurab, where discoveries of the highest importance in the history of architecture have been made, including the earliest

[-5-]

known colonnades.

Airplane Trip

The ride with Lord Allenby to Abu Roash had offered opportunity to explain to him the necessity of photographing the desert margin from an airplane, which might thus disclose prehistoric cemeteries, too faintly defined to be observable from the ground. At my request therefore Lord Allenby directed the commander of the Royal Air Force at Cairo to place a plane and pilot at my disposal for an experimental trip. On January 13th, 1920 I flew with this plane from the Heliopolis aerodrome across the southern delta to Abu Roash and then southward along the edge of the desert, traversing nearly the whole sixty-mile pyramid cemetery. I was told that a first flight is usually limited to twenty minutes, but in order to cover the ground it was necessary to stay up some two hours and circle repeatedly over the various sites. It was an extremely "lumpy" day, and I suffered greatly from sea-sickness. The lumpiness forced us to stay up about 5000 feet, and this seriously reduced the size of the negatives. I secured negatives of the leading pyramid cemeteries nevertheless, but my stay in Cairo was too limited to carry the experiment further, and I found myself far too busy to go on. The officers of the Royal Air Force however, understand what is needed, and have continued making negatives of the leading sites along the desert margin. A set of prints from these negatives will be placed at our disposal for filing in the archives of the Oriental Institute. It may be of interest to mention that the University of Chicago was the first institution to begin archaeological work from the air in Egypt.

Upper Egyptian Trip

The necessity of seeing more of the researches in Upper Egypt than the Abydos-Amarna trip with Mr. Greg had afforded, was one of the reasons why I was unable to continue the airplane work. On January 20th Mr. Bull and I left Cairo for Luxor, where we spent ten days. The Metropolitan Museum Expedition invited us to live at their comfortable expedition house on the west side of the river, and we spent part of our stay at Luxor as their guests.

As far as possible we examined all the newly cleared or restored tombs in the vast Theban Cemetery in the midst of which the Metropolitan Museum

Expedition have their house, besides other excavations of importance, especially the palace of Amenhotep III. We also examined the extensive evidences of the life of prehistoric man here, both on the plateau and in the valley below. The purchase of antiquities consumed a great deal of time, but this subject is discussed under Purchases below.

[-6-]

While at Luxor we were joined by two more members of the expedition, William F. Edgerton, Fellow of the Department of Oriental Languages, and Prof. A. W. Shelton of Emery [sic] University. With Mr. Bull and the two just mentioned I then had a group of three graduate students of the department, who were acquainted with oriental languages and able to make some study of the documents in the cemetery under my supervision. For a period unfortunately much too brief we held a very unusual seminar in the great Theban cemetery.

2. PURCHASES OF ANTIQUITIES AND WORKS OF ART

Examination of Collections

Wise application of the funds at the disposal of the Oriental Institute made it necessary to examine thoroughly from beginning to end all the private collections for sale and all the dealers' stocks available in Cairo. The latter were greatly congested because of accumulation during the continuance of the war, when European museums were no longer making their annual selections and the entire body of tourist travelers was also lacking. This work consumed a great deal of the time needed for scientific work at the museum, and all told was a matter of many weeks. See especially Andre Bircher Collection, below.

Purchases

A just account of actual purchases would require systematic exhibition of all the objects well installed and consideration of the exhibits to show how they have been built up out of various combined purchases. It is hoped that such an exhibit can be made if permanent space can be secured pending the departure of the Divinity School from Haskell Oriental Museum building. Under the circumstances only a few of the outstanding purchases can be mentioned, such as the following:

Complete group of 25 painted limestone mortuary statuettes representing the deceased and the members of his family including some 20 servants or children of the deceased engaged in all sorts of household activities like grinding flour, moulding and kneading loaves, cooking food, slaughtering cattle, or even playing the harp. They date from the Old Kingdom (3000 to 2500 B.C.) and form the most extensive group of such figures ever discovered in one tomb.

Group of royal seal cylinders including the official seal of Pharaoh Snefru, builder of the great pyramid of Dahshur; and the famous Queen Ahmose-Nofretere (see her bronze toilet mirror below).

Group of some 75 alabaster vases including ten inscribed with the names of various kings and queens.

[-7-]

Group of about 150 pre-Dynastic and Early Dynastic hard stone vases and other similar vessels. Several of the early examples are quite large and one is inscribed with the name of Pharaoh Aha-Menes, the first of the Pharaohs (about 3400 B.C.), the earliest known royal vase. About half of these were selected from the huge Andre Bircher Collection, numbering thousands of objects filling a native house in Cairo rented by Bircher for the purpose.

Series of 13 royal mortuary statuettes each inscribed with the name of a king or queen.

Group of about a hundred bronzes, including some 65 statuettes of which a number are of unusual size and some of very fine workmanship; a seated figure of Amon is adorned with golden jewelry and bears an inscribed dedication of Queen Shepenupet; two of the seated figures, a Sekhmet and an Imhotep are of silver-bronze (potin). Among four mirrors one bears on the handle the name of the famous Queen Ahmose-Nofretere, whose seal we also secured (see above). One of a series of battle-axes was that of an Egyptian army officer, with wooden handle and leather thong lashings still in perfect preservation since the Egyptian Empire (1580 to 12th Century B.C.). This group forms the finest collection of bronzes ever brought from the Near East to America.

Fine group of some 25 sculptor's model studies in limestone.

Official marriage announcement of Amenhotep III and his Queen Tiy, engraved on a large glazed scarab beetle. About 1400 B.C.

Beautifully written papyrus roll of the Book of the Dead, with black and white vignettes of unusual beauty and refinement. Probably of Saitic Date, of the 7th or 6th Century B.C. Far the best manuscript of this book as yet brought to America. The gift of Mrs. Elizabeth Milbank Anderson of Greenwich, Conn. (N.B. Negotiations are still going on as to whether Mrs. Anderson will contribute the entire cost of the papyrus or only half. It is hoped that she will contribute the whole cost and that it may be called in her honor Papyrus Milbank.). This manuscript is written in hieroglyphic and together with the hieratic copy from Paris (see above) gives us a fine example of both types of manuscript.

Group of some fifty glazed fayence statuettes and amulets.

The Timins Collection of stone weapons and implements. This series of over sixty fine pieces, together with a number of others found elsewhere, gives our Oriental

[-8-]

Institute the finest collection of Egyptian Stone Age industries in America.

A series of four variegated glass bottles in blue, white and yellow, representing the earliest stages of the glass vessel industry (14th Century B.C.).

Wooden statue of Theban noble, 2300 to 2000 B.C. 1/3 life size.

Beautifully painted mummiform coffin of 10th Century B.C.

Many historical documents in the form of reliefs and inscriptions on stone from the oldest period down to Greek times.

Series of 258 cuneiform tablets from Asia, presumably from Cappadocia, containing business records.

Large body of small objects for the study of Egyptian arts and crafts, making a considerable collection of the usual types.

Purchases for other institutions

The Chicago Art Institute placed $15,000 to the credit of the director of the Oriental Institute to be expended in sculptures for the Art Institute. A good deal of time was spent in making the proper selections. Just as he was leaving Cairo, and unfortunately too late to be used the director received $3,000 from the St. Louis Museum, with the request to expend it for Oriental antiquities.

3. RELATIONS WITH GOVERNMENTS & PREPARATIONS FOR ASIATIC TRIP

English

The cordiality of our relations with the English is well illustrated by Lord Allenby's hearty coöperation with my efforts to begin airplane photographic records. I was frequently asked to meet the Milner Commission to discuss Egyptian affairs. Besides meeting Lord Milner, I had numerous conferences with Mr. Alfred Spender, Editor of the Westminster Gazette and Secretary of the Milner Commission. At his request I furnished the Commission with a comprehensive plan for the reorganization of the Department of Antiquities of the Egyptian Government, much of which Mr. Spender has since informed me has been adopted and recommended in the report of the Milner Commission to the British Government.

Just before leaving Londin [sic] it was evident that our plans for our Asiatic expedition could not be put through

[-9-]

without support from the British Cabinet itself. I therefore wrote to Mr. Balfour a few days before my departure from London, explaining the situation and asking the cooperation of the London Foreign Office in our effort to begin scientific work in Western Asia. Shortly after arriving in Cairo I received a kind letter from Mr. Balfour stating that he was relinquishing the Foreign Office to Lord Curzon, but assuring me that he had recommended the support of our work to his successors. A letter from the Foreign Office soon assured me that Lord Curzon had written to Lord Allenby and the Cairo Foreign Office, as well as the Civil Commissioner in Mesopotamia instructing them to give us every necessary aid. From that time on I had only to apply to the Cairo Foreign Office, where my friend Mr. Robert Greg the Director-General, met our every want with the friendliest interest. Our chief difficulty, transportation to Mesopotamia by way of Bombay (as conditions made it impossible to go out there overland from the Mediterranean), was thus overcome.

French

The French Minister at Cairo, M. Lefevre-Pontalis is an old friend of Emile Senart, President of the Societe Asiatique. He at once showed a cordial interest in our enterprise and I was often at his beautiful residence. He supplied me with letters to the French provisional government at Beyrut, and a general letter also to all French officials whom we might meet on the frontiers of Asiatic territory in French occupation. He likewise informed the French Government at Beyrut of our proposed travels in those regions, received a favorable reply and handed me an official authorization to traverse French Syria.

Assembly of the Expedition

On Feb. 2nd, as we arrived in Cairo from the trip in Upper Egypt, we found Dr. D. D. Luckenbill of the Department of Oriental Languages, the last member of the expedition awaiting us there. For the first time the personnel of the expedition was then complete, including besides the present writer the following four gentlemen:

> Professor D. D. Luckenbill,
>
> Professor A. W. Shelton,
>
> Mr. Ludlow S. Bull,
>
> Mr. William F. Edgerton

By the seventeenth of February all was in readiness for the departure into Asia by way of Bombay.

III. ASIA

Sailing from Port Said on Feb. 18th, 1920, the party arrived without incident on Sunday, Feb. 29th in Bombay. After only forty-eight hours delay the party sailed on March

[-10-]

2nd for Basrah where they arrived on March 9th and disembarked on the 10th. They were met by the Chief-of-Staff from the Headquarters of the River Command, who took the director up to Headquarters to be the guest of the Commander, General Nepean; while the others were comfortably quartered at the hotel conducted by the military authorities. A staff car was at once placed at our disposal and in spite of the enormous extent of territory covered by the supply depots at Basrah, the car enabled us to assemble our supplies and equipment rapidly.

1. BABYLONIA

A few weeks before our arrival in Basrah the railway from Basrah up the Euphrates side of the alluvial plain to Baghdad had been completed. This railway was placed at our disposal and the University of Chicago expedition was the first archaeological expedition to use the Basrah-Baghdad railroad.

Lower Babylonia

Leaving Basrah by the night train on the 16th of March, with our supplies and equipment in a "goods van", we arrived at Ur Junction, some 120 miles from Basrah on the morning of the 17th. We were permitted to keep the railway van for the permanent safeguarding of our stuff, while we made excursions out from the railway to ancient sites we desired to study. After visiting Ur, and Eridu 16 miles south of it, we proceeded up the Shatt el-Hai, some eighty miles northward of the railway, through a very wild region over which had marched the expedition which had endeavored to succor General Townshend before his surrender to the Turks at Kut el-Amara. Besides the important Sumerian sites of Lagash and Yokha, which contain remains reaching back of 3000 B.C., we visited a number of unidentified city mounds on both sides of the Shatt el-Hai, a little-explored region which was evidently very thickly populated at an enormously remote date. We saw much of the admirable work being done by the British in civilizing this turbulent district of wild nomads who had not paid any taxes to the Turks for fifteen years before the war.

Returning to the railway at Ur we moved up the line through Lower Babylonia, making local trips away from the railway either in motor launches on the river (Euphrates) or in automobiles, all furnished by the British administration. In this way the remaining sites of Lower Babylonia were visited, especially Senkereh, Warka and Niffer, the scene of the work of the Philadelphia expedition.

Upper Babylonia

By March 29th we had reached Hillah six miles from the ruins of Babylon. Here General Wauchope was very kind and finally took in Mr. Luckenbill and myself as his guests.

[-11-]

We spent a large part of a week studying the ruins of Babylon, left just as the German excavations had uncovered them, and made a great many photographs, copies and plans. Beside Birs Nimrud we also visited Nejef, the sacred city of the tomb of Ali, Mohammed's son-in-law, which is 40 miles south of Hillah, and until the British conquest has been closed to non-Moslems with few exceptions. Before we left, General Wauchope invited a number of leading officers from G.H.Q. in Baghdad to meet him in Babylon and I had the pleasure of taking them through the ruins of the chief buildings. They were most interested in the Festival Street, the paving of which, laid by Nebuchadnezzar, must often have been trodden by the feet of the Hebrew exiles whom this mighty king carried away from Jerusalem.

Still having with us our "goods van" with the outfit and provisions, we arrived in Baghdad in the evening, April 5th. General Percy Hambro, the Quarter Master General kindly took me in as his guest, and the other members of the expedition were put up at the Hotel Maude. Finding that the railway north of Baghdad differs in gauge from the Basrah-Baghdad stretch we therefore relinquished the "goods van" and stored our stuff at the Officers' Hostel. Besides visiting some neighboring ruins, especially the marvelous palace hall at Ctesiphon, our time in Baghdad was chiefly spent in studying the tablets and other antiquities in the hands of dealers. At the same time many preparations for the trip up the Tigris across Assyria to Mosul (Nineveh) were necessary. Both General Hambro and Col. A. T. Wilson, the Civil Commissioner aided us without stint in all these preparations.

2. ASSYRIA

Assur

On April 12th all was in readiness for our northern journey up the Tigris by rail to Shergât, something over 180 miles by train from Baghdad. Shergât is still the rail head and likely to remain so for a long time. We were put up here at a rest camp while we studied the remarkable ruins of Assur, the earliest capital of Assyria, founded at least as early as 3000 B.C. The place had been completely excavated down to the primitive rock by the Germans and their work was finished before the outbreak of the war. It is the only site in Western Asia west of Troy which has been so completely investigated and it proved extremely instructive.

Mosul, Nineveh, Khorsabad and Nimrûd

Leaving Shergât by automobile on April 14th we made the run of some 80 miles along the Tigris, up to Mosul, where the commander, General Fraser very kindly took me in and arranged for the balance of the expedition to be put up at a native hotel. We began at once the study of the ruins of Nineveh, the latest Assyrian capital, lying across the

[-12-]

Tigris directly opposite Mosul. This kept us busy until an ebullition of the Kurds had settled down and we were permitted to run out about 15 miles northeast of Mosul to the foothills close under the northern mountains to visit the ruins of Khorsabad, the royal residence of Sargon II, father of Sennacherib. The palace has entirely disappeared since the French excavations, but in following the line of the extensive walls we found that the gates, of which there were several on each side, had been structures of monumental architecture. Though now grass-grown hills, each of these gates could be excavated with great profit, for the natives uncovered in our presence a large alabaster threshold in one of the gates, bearing impressive cuneiform records of the campaigns of Sargon, the conqueror of Samaria, and the Assyrian king who carried away the captives of the Ten Tribes of Israel and exiled them.

Crossing the river to the east side, we were also able to move down the Tigris 20 miles below Mosul, to the second capital of Assyria, the Biblical Calah, now called Nimrud (pronounced Nimrood). The temple tower and the palaces here are in an unusually good state of preservation. Many sculptures and inscribed records project from the encumbering rubbish, ensuring magnificent returns for excavation, and a great opportunity for recovering and reconstructing an entire Assyrian city as well as a tremendous chapter of human history. We were accompanied in our inspection by the owner of the land occupied by these ruins and gained his friendship. We accepted his invitation to dine at his house as we were returning to Mosul, and found it was near the ruins of Balawat, an Assyrian palace of the 9th Century B.C., which we also saw. It was from this palace that Rassam many years ago took out the massive bronze mountings of a palace gate richly adorned in repousse designs. Nothing has since been done there.

We had now traversed the Tigris going up-stream, to the region where it issues from the northern mountains. North of us was a Kurdish population quite unsafe to penetrate. Indeed, the whole Mosul region was a hazardous one. A few days before our arrival a British officer was murdered by the ruins of Assur. Of the fifteen political officers of the British administration, seven were murdered by natives, five before our arrival and two afterward. Such unsafe conditions are however, evidently only temporary.

Having ascended the Tigris about 275 miles above Baghdad and some 625 miles from the Persian Gulf, our return to railhead at Shergat was delayed by a terrible cloud-burst storm which washed out the bridges. When we finally reached Shergat again on April 20th we found the railway broken in two places by the storm, while hostile Arabs had cut it in a third place. We were completely cut off from Baghdad and unable to reach it again until April 23rd.

[-13-]

5. OVERLAND RETURN JOURNEY ACROSS THE NEW ARAB STATE TO ALEPPO

Arrangements in Baghdad

On returning to Baghdad the Civil Commissioner informed me of the discovery of a series of remarkable ancient wall paintings uncovered during the excavation of a rifle-pit in the enormous Roman stronghold of Salihiyah occupied by the British as their furthest outpost on the upper Euphrates some 300 miles above Baghdad. He asked me to go there at once and make a record of the paintings and a series of photographs that they might not perish and be lost to modern knowledge. As the British authorities had thus far feared to allow the expedition to go up the Euphrates more than at most a hundred miles because the region was still a fighting zone, I seized the opportunity with the greatest pleasure; but asked for a fortnight to be spent among the monuments on the Persian border first.

The Civil Commissioner then stated that if we went to Persia first we would be too late to save the paintings, for the reason, then strictly confidential and known only to the High Command, that the British frontier on the Upper Euphrates (toward Syria and Faisal's kingdom), was to be drawn in about a hundred miles further down river, because of excessive difficulties in such a long line of transport communications. If we went to Persia first the paintings would lie out a hundred miles beyond the British lines, and equally far in Arab territory, that is they would be quite inaccessible on our return from Persia. It was evident that we should leave for the Upper Euphrates at once.

I then asked the Civil Commissioner why it would not be possible, on completing our work at Salihiyah, to proceed up the Euphrates and go on to Aleppo, and thus return to the Mediterranean overland instead of coming back to Baghdad and repeating the long return voyage via India and across the Indian Ocean to the Red Sea and the Mediterranean. He replied that there was of course great risk, but that the probabilities were in our favor, as the Arabs would be in a genial frame of mind as a result of having recovered so much of the Euphrates valley. I then asked the Civil Commissioner to telegraph to Salihiyah to Col. Leachman, who had traversed the region several times in former years and had long been acquainted with the sheiks of the tribes through which we would pass on our way to Aleppo, and to ask his opinion. Col. Leachman replied the next morning stating it was "probable" the Chicago expedition could get through. The Civil Commissioner then agreed to furnish two of the seven automobiles we needed, provided the Commander-in-Chief in Mesopotamia, General Haldane, would furnish the other five cars and give us permission to undertake the trip at all. I took advantage of a lunch with General Haldane to bring up these matters, and thus secured the automobiles and the needed permission as well.

[-14-]

From Baghdad to the British Frontier on the Upper Euphrates

On Wednesday morning, April 28th our seven automobiles crossed the Tigris and swung out of the southern suburbs of Baghdad and drove straight west on the first lap of the overland journey to the Mediterranean. Having crossed the Euphrates above Falujah the journey was made on the right (south or west) bank of the river. It was planned that we should arrive each night at a British post, but frequent accidents and delays forced us to stop short and two nights were spent unprotected in the open desert with Bedwin campfires visible all about us. The British officials showed great anxiety, though we saw no signs of danger. A few weeks later however, Col. Leachman, above referred to, was murdered by the Arabs in the same district where we spent our first night in the open desert, near Falujah.

Accidents, breakages and delays of desert travel were such that the three-hundred mile trip to the British frontier occupied an entire week. The last day or two we were convoyed as we were passing points which were often under Arab fire. General Cunningham in command at Salihiyah received us most kindly, and as his quarters were entirely full, Col. Leachman had us set up our field beds in his <u>office</u>. Every possible kindness was shown us by the British officers along the entire trip. General Cunningham sent Mr. Luckenbill and myself for an air reconnaissance in one of his bombing planes, an experience which gave us exceedingly valuable impressions of the desert and the Euphrates valley.

The British withdrawal from Salihiyah down the Euphrates was begun on the fifth of May; this left us only the fourth on which to make our records of the paintings, which proved to be of unusual interest and value. The British officer in command of the spot, Major Wrightwarren, placed a squad of Indian troops under a sergeant at my disposal to shift sand-bags in order to lift the camera to the proper level. Mr. Luckenbill made 24 negatives of the paintings and the ancient sanctuary containing them. The young men made a ground plan of the structure, while the director spent the day in making as full notes as possible on the paintings and inscriptions. I then suggested to the major that the squad he had given us might be set to work covering the paintings with rubbish again and thus protecting them from destruction by the Arabs. He at once gave orders that this be done.

From the British Frontier on the Upper Euphrates to Aleppo

As the British retired down river and we were to continue our journey up the Euphrates, it was of course necessary to surrender our seven automobiles to General Cunningham. On the morning of May 5th we shifted to five native wagons or "arabanahs" and we drove in these out of

[-15-]

the north gate of the ancient fortress of Salihiyah before dawn as the British were preparing to withdraw through the south gate. By the good offices of Col. Leachman five Arab rifles of a neighboring friendly sheik met us as we drove away and escorted us over no man's land into Arab territory. We thus left British and committed ourselves with much misgiving to Arab protection. In a few hours we were met by five other Arab horsemen, sent by the Arab government of King Faisal from Der ez-Zor to meet us and relieve the local rifles who had first met us.

The journey from the British frontier up the Euphrates and across from it to Aleppo occupied a week. It was an anxious, rough and difficult week. The Arabs showed the greatest friendliness toward us as Americans; had we been British on the other hand, or French, we never would have come through alive. We had much opportunity to meet the sheiks and tribesmen and their hatred of British and French was intense. I found it at first difficult to believe that the traditional Arab friendship for the English had been displaced by bitter hostility; but many striking experiences revealed the change. A deputation of officers of the Arab army called on me at Der ez-Zor to send messages imploring assistance and advice from America. The seriousness with which they voiced their resentment toward England, their need of guidance and advice, and their earnest desire for assistance from America was very impressive. Their friendliness was very appealing. They were ready to give us all protection, and our chief danger lay in the roving bands of brigands infesting the country. On May 12th we rode safely into Aleppo.

4. SYRIA

We had hoped that it would be possible to penetrate south-eastern Asia Minor from Aleppo but found this unfortunately quite impossible. The Arabs hovering on the flanks of the French threatened to cut the railway south from Aleppo, and we were urged to leave for Beyrut as quickly as possible. The conditions throughout Syria were very unfavorable for carrying out the archaeological reconnaissance which we had hoped to make.

From Aleppo to Beyrut

It was however very important that we should inspect the ruins at Kadesh and Baalbek as [we] went south. I secured a letter from the Arab governor of Aleppo to the local authorities in the Orontes Valley, who furnished us with escorts, and we were thus able at considerable risk to inspect the important ruins at the two leading points between the Lebanons, Kadesh and Baalbek. On the 18th of May we reached Beyrut.

[-16-]

The Phoenician Coast

Dr. H. H. Nelson, head of the History Department at the American College in Beyrut, and doctor of the Department of Oriental Languages at Chicago, gave us a warm welcome and was of the greatest assistance to us in exploring the Phoenician coast. The institution gave him complete freedom from duty so that he could accompany us everywhere, and he became temporarily a member of the expedition. In motor cars we explored the Phoenician coast northward from Beyrut as far as some twenty miles north of Tripoli, that is to the northern end of Lebanon, where we were stopped by the depradations of brigands whom the French were powerless to control.

Going southward from Beyrut to reach Tyre and Sidon in the same way, I found the French authorities most friendly, as they had been notified of our coming, and cordially responded to all requests for protection or assistance; but as we were about to leave Sidon and push on southward to Tyre, news came in that three men had just been shot by brigands a few miles out on this road, and the French commandant urged us to turn back. We were quite willing to comply.

At Sidon we were entertained at lunch by Dr. George A. Ford, of the American Mission, who showed us some examples of his extraordinary Phoenician collection,- especially the sculptured sarcophagi- which he wishes to dispose of for the benefit of his orphanage school. This is an opportunity to secure the best Phoenician collection ever made.

While the turbulent conditions limited the extent of our Phoenician survey very disappointingly, nevertheless we secured a great many archaeological and topographical data of much value, and many photographs. Besides a very

satisfactory conference with M. Chamonard who is in charge of the French Service des Antiquites at Beyrut, I also had an interview with General Gouraud,
the French High Commissioner governing Syria. I am confident that any future
archaeological work by our Oriental Institute in Syria will meet with cordial
French support.

Beyrut to Damascus

The journey by railway from Beyrut to Damascus was without incident,
but the stay in Damascus was very profitable and interesting. A letter from
Lord Allenby to King Faisal procured me an interview with the new Arab ruler,
and I afterward dined with the king in company with the American Consul. I
learned much of value for our future relations with this region in the continuance of the work of the Oriental Institute. Among these experiences was a
session of the new Syrian Parliament, and a long conference with the President of this body who called on us at the hotel. Two members of King Faisal's
Cabinet are graduates of the

[-17-]

American College at Beyrut, and besides these gentlemen we met a number of
other educated Syrians who are members of the Parliament, and we listened
with the greatest interest to their debates.

5. PALESTINE

Damascus to Haifa

This journey by rail was directly across a disaffected region south of
the Sea of Galilee where we saw a brigand hanging from a telegraph pole beside the railway line. From Haifa we explored the north side of the Plain of
Megiddo, which was likewise rather unsafe. A stupid guide mislead us so that
we failed to reach Megiddo itself, although we could see the great mound a
few miles away across the plain, and discern what great opportunities for
excavation still await the investigator there. We here had opportunity of
studying the earliest great battlefield between Egypt and Asia,- the scene of
so many dramatic struggles between the nations that it has become proverbial
as Armageddon.

Haifa to Jerusalem

At Haifa Messrs. Luckenbill and Nelson turned back to Beyrut, for it had
now become evident that our projected summer of exploration in Syria and Palestine would be quite impossible in view of the turbulent conditions. At Beyrut Mr. Luckenbill busied himself developing our great body of photographic
exposures, which it was not safe to bring back to America and expose them to
a sea voyage before developing. The director and the remainder of the party
went on to Jerusalem. I had a series of very valuable conferences at Jerusalem with the British authorities, especially with Sir Louis Bols, Commander-
in-Chief of the British Army in Palestine; Prof. John Garstang, Director of

the British School of Archaeology in Jerusalem; and Captain Mackay whom the British have placed at the head of their service for conservation of the ancient monuments. But even around Jerusalem the country was so unsafe that it was impossible to go out and inspect a ruin as near as the mound of Jericho in the Jordan valley, and practically visible from the Mount of Olives.

The conditions in Palestine were disquieting for a number of reasons. Both Arabs and Christians, that is some 90 per cent of the population, were bitterly discontented with British control, because they said they had been promised British rule (which they welcomed), whereas they had been given Jewish rule. I had seen enough of the conditions among the Arabs on the east and north of Palestine to realize that a Bolshevist wedge pushing southward through the Caucasus with the aid of the Turkish Nationalists, would meet with a very hospitable reception among the Arabs and without

[-18-]

doubt also from the Jews of Palestine, many of whom are already infected with Bolshevism. When I told Sir Louis Bols of these things, I found him deeply concerned.

Jerusalem to Cairo

This journey for the first time in my experience, was now possible by rail, following the line of march of armies between Africa and Asia for five thousand years. I went with General Waters-Taylor, Head of the Intelligence Department of the Imperial Staff. This offered [the] opportunity of spending the entire day in conversation with one of the best informed men in British service regarding Western Asia. He was going to Cairo to consult Lord Allenby.

6. PURCHASES OF ANTIQUITIES IN WESTERN ASIA - OTHER RESULTS

The journey from Jerusalem to Cairo on June 3rd completed the work of the expedition in Asia, and the following paragraphs furnish a brief summary of the results.

Purchases

The most important purchase is a copy of the Royal Annals of Sennacherib. In form the document is a six-sided prism of buff colored terra cotta, or baked clay, hard and firm and in perfect preservation. Six columns of beautifully written cuneiform fill the six faces of the prism. In content it records the great campaigns of the famous Assyrian emperor, including the western expedition against Jerusalem on which he lost a great part of his army, - a deliverance for the Hebrews which forms the supreme event in the life of the great statesman-prophet Isaiah. It is a partial duplicate of the Taylor Prism in the British Museum, but ours is older having been written three years earlier under another Eponym. The nature, extent and value of the variants can only be determined by an exhaustive comparison. No such monument as this has yet been acquired by American museums, and besides its scientific

usefulness it will form an exhibit of primary value to students and of unique
interest to the public.

Of other cuneiform documents our purchases total nearly if not quite a
thousand tablets of varying content, including some that are literary and religious. Among works of art, besides two early Babylonian statuettes of cop-
per, a very fine series of some forty beautifully cut stone cylinder seals,
of which the best is one of the finest examples of lapidary sculpture ever
found in Babylonia. A few examples of the composite Syrian lapidary art were
also secured at Aleppo. On reaching Sidon our funds were too depleted to undertake the purchase of the

[-19-]

magnificent Phoenician collection of Dr. Ford, and this remains one of several unique acquisitions which urgently need consummation.

It ought to be mentioned here that without excavation it is impossible
to gather by purchase in Western Asia collections of the wide range and remarkable volume possible in Egypt. To expand our Asiatic collections, excavation will be necessary.

Other Results

Not least among the valuable results of the Asiatic expedition was the
acquaintance with the archaeological remains, the geography and topography of
Western Asia gained by the members of the expedition. This knowledge is re-
inforced by a very large and complete series of photographs, and extensive
field notes, including many plans and descriptions. An extensive series of
maps, plans and diagrams exhibiting the geography, topography and ethnology
of Western Asia prepared by the British authorities has also been acquired.

The facts regarding prices of labor, the season when labor is free to
leave flocks and fields, the possibilities for disposing of excavated rubbish, and all items of information essential to carrying on excavations at
all important points in Mesopotamia, Syria and Palestine were carefully collected.

The question of personal and official relations with controlling authorities was also given careful attention. We made the acquaintance of many
officials of England and France now permanently stationed in the Near Orient, and as far as the regulations have been formulated we learned the conditions under which future work of excavation may be carried on in territory
now controled by the two powers mentioned. The Civil Commissioner at Baghdad
showed me his entire file of official records concerning excavations and the
status of ancient monuments. He assured me he would welcome an expedition of
the University of Chicago, which might desire to excavate in Mesopotamia, and
that we could count upon having the site of Nimrud (see above) if we desired
it. He asked me to draught an outline of the best organization for a Department of Antiquities for the Mesopotamian Government. At the same time, Maj.
Bowman, Director of the Department of Education, who is temporarily in charge

of such matters, also asked me for such a draught, and I still have this task before me. (See below, VI. OPPORTUNITIES & RECOMMENDATIONS)

We also established connections with a number of sheiks and natives of influence, whose assistance would be indispensable in undertaking field work in Mesopotamia.

[-20-]

Our relations with the authorities in Palestine are similarly favorable, should we be able to undertake work there. Prof. Garstang, Director of the British School in Jerusalem voluntarily offered to hold in reserve for the University of Chicago, the splendid fortress city of Megiddo (see above), in case we should be able to undertake excavations in Palestine.

IV. POLITICAL MISSION TO ENGLAND

1. CAIRO EVENTS

Whether General Waters-Taylor, Chief of the Intelligence Department, with whom I journeyed from Jerusalem to Cairo had informed Lord Allenby in advance regarding the overland journey of our expedition from Baghdad to the Mediterranean or not, I do not know; but on my arrival in Cairo, as I was paying my respects to Allenby at the Residency on the first day of my return, he invited me to luncheon with a considerable number of the leading British officials of the Near East. Allenby took this opportunity to bring up the situation in Western Asia for general discussion, and asked me to go to England to report to the British Government the facts which had come under our observation. Although I had already engaged passage to America via Naples to New York, Lord Allenby assured me his secretaries would dispose of these tickets, and he would secure me passage to England on the same ship with Lady Allenby, then just returning to England for the summer. He left the company while he went and wrote me a letter to Mr. Lloyd-George and another to Earl Curzon, British Foreign Minister. In a few days his secretaries furnished me with a laisser-passer, and diplomatic visas, besides all the necessary tickets, and a copy of Allenby's long cablegram to the British Government, in which they were asked to reimburse me for any expenses incurred over and above those involved in the voyage via Naples.

2. CAIRO TO ENGLAND

Lord Allenby kindly invited me to a drawingroom [sic] in the special train in which he accompanied Lady Allenby from Cairo to the ship at Port Said, and on the fifteenth of June we sailed for Plymouth, arriving without incident on June 26th, and reaching London the same day.

3. REPORT TO THE BRITISH MINISTERS IN LONDON

Immediately on my arrival the Spa Conference called the Prime Minister away and I did not see him. He left instructions that the Minister for India was to receive me and take my report. With Earl Curzon I had a long conference at the Foreign Office. There were three main points in the report which I gave him:

[-21-]

First. The dangerous hostility of the Arabs and the threatened general outbreak against the British. I was thus able to forewarn him of the imminent Arab outbreaks which have filled the press dispatches during the last few weeks. Indian troops were at once dispatched to Mesopotamia.

Second. The dangerous situation in Palestine resulting from the disproportionate amount of power granted the Jews, who form but ten per cent of the population of the country.

Third. The persistent anti-British propaganda carried on by French officials, and the reprisals in kind by the British, introducing into Western Asia a European rivalry which has already had deplorable consequences in Syria and Palestine, as well as in Mesopotamia.

Lord Curzon was very cordial in his thanks, and I received also a letter of thanks from Lord Hardinge of Penshurst enclosing a check for all extra expenses incurred by changing my homeward route.

V. SECOND TRIP TO PARIS

1. NEW PURCHASES OF BABYLONIAN AND ASSYRIAN RECORDS

While in Baghdad I accidentally learned of two extraordinary pieces which had been sent by Baghdad owners to obscure Paris dealers for sale. I took advantage of the journey to England therefore to run over to Paris for a few hours and succeeded with some difficulty in locating these pieces.

The first is a small tablet of pure gold engraved on both sides with a cuneiform record of the restoration of one of the early temples of Assur by Shalmaneser III (859-825 B.C.) accompanied by a summary of his great wars. It was deposited under a large slab of stone beneath the Holy of Holies of the temple of Ishtar at Assur.

The second is a crescent shaped portion of a tiara or collar of Shargali, king of old Babylonia about 2700 B.C. It is also of solid gold and is engraved with the king's name and dedication in cuneiform. The Paris dealer allowed me to take the piece on orders from the Baghdad owner, with whom however it has thus far been impossible to arrange a satisfactory price.[2]

A group of important cuneiform records including royal annals of the Chaldean Age, and five archaic tablets with picture writing from which the cuneiform grew up.

---

[2] This piece was eventually returned to the dealer [ed.].

2. LORD WIMBORNE'S ASSYRIAN SCULPTURES

Among the early English explorers who brought back oriental monuments was Lord Wimborne, who installed on his estate in England in a building especially erected for the purpose, a large series of Assyrian palace wall sculptures.

[-22-]

Among them are two huge winged bulls over ten feet high carved in alabaster,- the sentinel animals guarding the entrance of the palace. These are the creatures adopted by the Hebrews and called "cherubim" which is their Assyrian name. On this hasty visit in Paris I met one of the Armenian Kelekian Brothers, wealthy antiquity dealers, who informed me that he had purchased these Assyrian sculptures from the present Lord Wimborne who has no interest in them. He stated that he had shipped them to New York where they now are in storage, expecting to dispose of them to some American museum. This purchase is a very unusual opportunity.

VI. OPPORTUNITIES & RECOMMENDATIONS

PURCHASES RECOMMENDED

The outstanding opportunity is the magnificent group of Lord Wimborne's Assyrian sculptures just mentioned. No such opportunity will ever occur again, as it is the only private collection existent, which contains such an extraordinary body of Assyrian monumental sculpture. I have been informed on good authority that the Metropolitan Museum is too deeply involved in Egypt to consider this purchase; Philadelphia and Boston are not likely to have the funds; Kelekian's purpose of involving the American museums in a bidding contest is therefore likely to fail, and the opportunity promises to be a favorable one.

The Phoenician collection of Dr. George A. Ford at Sidon is the most extensive existent group of such materials. It includes an extraordinary series of tomb sculptures, chiefly in the form of anthropoid sarcophagi, one of which, a specimen of beautiful Hellenistic-Oriental sculpture, is alone worth the money he asks for the whole collection. Among the sculptures is also a very rare piece, one of the kneeling horses forming the capital of a Persian column, which must have belonged to some Persian building in Sidon. Dr. Ford asks $25,000 for this collection, as a contribution to his orphanage at Sidon. The mingling of business and philanthropy in the transaction can in no way prejudice the fact that the collection is very cheap at this price.

A remarkable group of royal decrees inscribed on stone and supplementing a series discovered by the French at Coptos, is in the hands of the Christian Moharb Todros at Luxor, Egypt. They date from the 25th Century B. C. and belong among the few royal archives surviving from the Pyramid Age in Egypt. Todros asks 1500 pounds for them, but they can probably be had for half that sum. As historical documents such as are possessed by no museum in Europe besides the Louvre, they would form an accession of the first rank for our collections. Our funds were too low to undertake their purchase before I left Egypt.

[-23-]

2. OPPORTUNITIES FOR EXCAVATION AND RESEARCH

While wise purchasing will save much for science and bring it into our American collections, such buying can never do more than form part of a general plan for meeting the situation as a whole. The Near East is a vast treasury of perishing human records, the recovery and study of which demand a comprehensive plan of attack as well organized and developed as the investigation of the skies by our impressive group of observatories, or of disease by our numerous laboratories of biology and medicine. The fast perishing records demand a far reaching attack directly on the mounds covering the ancient cities and cemeteries, whence the natives by illicit digging which destroys as much as it brings forth, commonly draw the antiquities which they offer for sale. Furthermore, the ancient city itself with its streets, buildings, walls, gates, water-works, drains and sanitary arrangements is a fascinating and instructive record of human progress and achievement, which must be studied, surveyed and recorded, in the same way the geology, botany and zoology of the Near East must be studied to reveal the character of the habitat and resources of the earliest civilized communities of men.

These proposals will endeavor to indicate:

1. How and where a properly correlated group of expeditions might be set to work and developed.

2. How such a group of expeditions might be organized to become the contributing agencies furnishing all the surviving facts and sources for classification, filing, arrangement and study at a properly equipped headquarters in the Near East, which should become the central laboratory for the systematic investigation of the whole range of the human career throughout the entire Near East.

As already suggested, the most practical and tangible line of initial development would be a series of excavating expeditions at the most promising accessible sites, whether ruined cities or cemeteries. Taken up by countries, our expedition has shown the most promising places to be the following:

Egypt

The greatest royal cemetery in the world is at Memphis. Its tombs of kings and nobles contain remains which reflect the entire range of ancient civilization at the early period when its currents were beginning to set toward prehistoric Europe, then in Stone Age barbarism. This vast cemetery has thus far only been nibbled at. The Egyptian Government, while reserving it for excavation by its own Department of Antiquities, has in vain endeavored to meet the obligation thus assumed.

By actual computation by one of its own staff, it will take the Egyptian Government <u>five hundred years</u> to complete the excavation of the Memphite cemetery at the present rate of progress. Conferences with the Milner Commission gave me an opportunity to put this situation clearly before them and they concluded that excavation by the Egyptian Government if not discontinued should at all events he discountenanced in favor of a policy of yielding to private initiative the chief responsibility for rescuing such enormous bodies of records for scientific use.

A proposition to carry on the clearance of this unrivaled treasury of ancient human life on a scale commensurate with its size and importance, with a big and competent scientific staff, would eventually meet with acceptance and thus place in our hands the greatest and most important body of surviving sources of ancient civilization at its incipient stage.

Phoenicia & the Movement of Civilization toward Asia and Europe

The cemetery of Memphis has already revealed the earliest surviving representations of sea-going ships (28th Century B.C.) and their traffic with the Asiatic coast later known as Phoenicia. The Phoenician port of Byblos already in use by the Egyptians by 3000 B.C., we visited and found it lying ready for excavation. As the Memphite cemetery would furnish the source, so this coast would furnish the destination of the earliest civilization that ever arose. This illustrates the necessity of a group of expeditions and the correlation of their results.

The extensive cemetery of Sidon, which yielded the magnificent sarcophagus of Alexander (so-called) has only been very incompletely investigated. Much of this cemetery is on the land owned by Dr. Ford, and he assured me he would be very glad to have us complete the clearance of the tombs on his property. We inspected the ground and found it a very promising opportunity.

Assyria and Babylonia

At the same time the remains of early civilization in the hinterland of Western Asia, in the land of the Two Rivers (Tigris & Euphrates) offer opportunities not less great and important.

I have seen the records of Sargon, the captor of the Ten Tribes of Israel, buried in the gates of his city of Khorsabad. I have seen the monuments of many great Kings of Assyria preceding Sargon lying at Nimrud (the predecessor of Nineveh), where they project from the ground revealing the presence of a great Assyrian capital city which may still be recovered and planned. The British Civil Commissioner at Baghdad assured me orally that he would be glad to see us working at Nimrud.

[-25-]

A programme including the excavation of the gates of Khorsabad and the whole of Nimrud, could be carried out in a few seasons. This would offer opportunity for close coöperation with the British, and the further cultivation of the cordial relations with them, which we already enjoy. Meantime the British would find it increasingly difficult to ignore the hard fact that they are without either the men or the funds to excavate the great city of Nineveh. The reservation of Nineveh by the British for the last two generations has created a great obligation to science which they have thus far met by nothing more than a little haphazard grubbing.

Although they are still reserving Nineveh for themselves, the British are already anxious to make some form of combination with Americans in order to eke out their own meagre resources. A few successful seasons by an American expedition at the gates of Khorsabad and the palaces of Nimrud, would enable us to put the British in a situation where something would have to be done by them to make the records buried in the great Assyrian capital accessible to science. Diplomatic handling of this situation, would in my judgment put our expedition in command at Nineveh, where a decade of successful work would enable us to restore to modern knowledge the vast treasury of human records and human handiwork now lying buried in the greatest imperial capital of Western Asia. No task more pressing nor more illustrious in its achievement than the excavation and recovery of this magnificent Rome of Western Asia is to be found in the whole range of humanistic research.

At the same time we should be very advantageously placed for sending out and maintaining small branch expeditions to a number of important city mounds in Babylonia, which would reveal to us the earlier history of civilization along the Two Rivers, reaching back into the centuries preceding 3000 B.C.

<u>Hittites, Syria & the Western Movement of Babylonian Civilization</u>
<u>toward Europe.</u>

Babylonian influences found their way to Europe chiefly through the Hittites of Asia Minor, whose baffling inscriptions are only now beginning to be deciphered. The overflow of Hittite civilization into Syria has left a great landmark at Kadesh on the Orontes in northern Syria in the form of a great city mound of impressive height and length. We visited this imposing ruin and it is evident that its clearance would furnish a wonderful revelation of the composite civilization made up of Babylonian, Hittite and Egyptian elements, which developed in Syria and found its way to Europe through Asia Minor by land, and the ports of Phoenicia by sea.

[-26-]

<u>Palestine</u>

Similar common ground for the mingling of the great civilizations of Egypt, Babylonia and the Hittites is found in Palestine, with the added interest and importance due to the fact that it was the birth place of the greatest of religions. The strongest and strategically most important for-tress-city of Palestine was Megiddo, now famous under its later name Armaged-don, on the northern slopes of the Carmel ridge. The British victory under Lord Allenby, which restored Palestine again to Christian rule, was won at this place, the last of the long succession of inter-continental struggles on this oldest known battle-field of history. The British authorities have as-sured me that they will reserve this place for excavation by the University of Chicago.

Headquarters in the Near East for the Study and Correlation of Documents Excavated

The newly discovered documents and great groups of new facts thus brought to light by the excavations proposed above, would require classifica-tion and coördination and would therefore have to be gathered together at a common center where the process of study, correlation and publication could be steadily carried on. For this purpose there would eventually be necessary a winter headquarters at Cairo and a summer headquarters on the high cool slopes of Lebanon overlooking Beyrut. These two centers would together form an ORIENTAL INSTITUTE HEADQUARTERS on the ground and together constitute a common center furnishing both administrative and investigative direction of the work throughout the Near East.

The main objects of the Oriental institute Headquarters might be summa-rized thus:

1. The general administrative oversight and management of a group of lo-cal expeditions working among the remains of all the leading ancient civili-zations of the Near Orient, being chiefly the regions surrounding the eastern end of the Mediterranean Sea.

2. To furnish investigative direction and working quarters for a group of investigators who should receive, classify, correlate, study and publish the facts and sources discovered in the field in order to disclose and trace especially:

    a. The earliest evidences of man in the geological ages and his rise from Stone Age savagery to civilization.

    b. The development of the earliest civilized communities, espe-cially in government, business, city-building, art, architecture, literature and religion.

    c. The discovery of barbarian Europe by oriental civilization and the transplanting of oriental civilization to Europe.

[-27-]

    d. The culmination of Oriental civilization in the lofty religious vision of the Hebrews and its supreme expression in the life of Jesus.

e. The later relations of the Orient with Europe, culminating in the conquest of Europe by Christianity, an oriental religion.

f. On the basis of the above investigations, <u>to produce a work on "The Origins and Early History of Civilization", which shall give the first adequate account of human beginnings & the early career of man.</u>

This list of periods and subjects discloses at once the unique importance of the Near East in human history. It was not only the earliest home and source of civilization, which first brought civilization to Europe, but it was also the cradle of the supreme religion of today, besides other leading religions like Zoroastrianism, Judaism and Islam. This report, already too long, does not offer sufficient space to demonstrate the overshadowing importance of the Near East in the field of humanistic research at the present day. For this demonstration I must refer to my Convocation Address of September 3rd, 1920 (accompanying this report), which I wish might become a part of this report. From this address I would like to quote the following paragraph:

"Before the whole recoverable story drawn out of every available mound is in our hands, it may indeed be a century or two; but after a survey of most of the important buried cities of the Near Orient, I am confident that with sufficient funds and adequate personnel <u>it will be possible in the next twenty-five or thirty years, or let us say within a generation, to clear up the leading ancient cities of Western Asia and to recover and preserve for future study the vast body of human records which they contain.</u> In this way the main lines of the development can be followed in the larger sites, marking the leading homes of ancient men and governments. I cannot but see in the recovery and study of this incomparable body of evidence America's greatest opportunity in humanistic research and discovery".

To this statement from the Convocation Address, I can only add a reference to the impoverishment of European governments and their entire lack of men to do this work, as these facts are set forth in the address. This complete paralysis of Europe in oriental research thus not only shifts a grave responsibility upon the shoulders of America, but at the same time enlarges our own opportunity as never before.

In conclusion, Mr. President, I want to express to you, and through you also to the Board of Trustees, my deep appreciation of the great opportunity offered to my associates and myself in this preliminary journey of reconnaissance in the Near East, and my earnest hope that it may bring great results both for science and for the University of Chicago.

Very respectfully yours,

Director

APPENDIX C

BIOGRAPHICAL SKETCHES OF EXPEDITION PARTICIPANTS

JOHN A. LARSON

James Henry Breasted (1865–1935), American Egyptologist, Orientalist, and historian, was born in Rockford, Illinois, on August 27, 1865, the third child and elder son of Charles Breasted and his wife Harriet (Garrison). In the summer of 1873, the Breasted family moved to Downers Grove, Illinois, where James grew up and attended public school. By 1880, he began to take classes sporadically at North-Western [now North Central] College in Naperville, Illinois, where he eventually received a Bachelor of Arts degree in 1890. In the meantime, Breasted worked as a clerk in local drugstores and, in 1882, entered the Chicago College of Pharmacy, where he graduated in 1886. He then was employed as a professional pharmacist and acquired much knowledge about drugs, which was to prove useful in later life when he was dealing with ancient Egyptian medical texts. In 1887, Breasted began his study of Hebrew and Greek at the Chicago Theological Seminary, and subsequently was enrolled at Yale University in New Haven, Connecticut, in 1890/1891, where he was awarded a Master of Arts degree, *in absentia*, in 1892.

With the encouragement of William Rainey Harper, then Professor of Hebrew at Yale University, Breasted went to Berlin in 1891 to study Egyptology with Professor Adolf Erman who himself was a student of the pioneering German Egyptologist Richard Lepsius. James Henry Breasted became the first American to earn a PhD in Egyptology (University of Berlin, August 15, 1894) and the first to receive an appointment to teach the subject in an American university (University of Chicago: Assistant in Egyptology and Assistant Director of the Haskell Oriental Museum, from October 1, 1894 to 1901; Instructor in Egyptology and Semitic languages, 1896; Assistant Professor, 1898; Director of the Haskell Oriental Museum, 1901–1935; Professor of Egyptology and Oriental History, 1905–1935). His first appointment at the University of Chicago began with a six-month leave of absence, during which time he was scheduled to do "exploration work" in Egypt.

On October 22, 1894, Breasted married Frances Hart (1872–1934), a 21-year-old American student, whom he had met in Berlin. The Breasteds would eventually have two sons, Charles and James Jr., and a daughter, Astrid (the "little girl" of the 1919–1920 home-letters). The newlyweds spent a working honeymoon in Egypt during the winter of 1894/95, and Breasted acquired several thousand Egyptian antiquities for the new Haskell Oriental Museum (since 1931, the Oriental Institute Museum) at the University of Chicago.

During the next twenty-five years, the publication of a series of textbooks and technical works established James Henry Breasted as one of the senior Orientalists in the United States. From 1900 to 1904 he collected data for the great Berlin *Wörterbuch der Ägyptischen Sprache*, and the German academies in Berlin, Leipzig, Munich, and Göttingen asked him to copy and arrange hieroglyphic inscriptions in their collections. During the same period, he began work on the most important ancient Egyptian historical texts, including many unpublished ones, with the intention of producing a sourcebook of English translations for the benefit of historians in general; the accumulated 10,000 manuscript pages of translations and commentary were published in five volumes as *Ancient Records of Egypt: Historical Documents from the Earliest Times to the Persian Conquest* (Chicago: The University of Chicago Press, 1906–1907). This major corpus of primary source material enabled the ancient Egyptians to speak for themselves and served as the basis for Breasted's popular book, *A History of Egypt from the Earliest Times down to the Persian Conquest* (New York: Scribners, 1905), in which he drew his conclusions from his translations of the ancient texts.

For two winter seasons, 1905–1907, Breasted was director of an epigraphic expedition to Egypt and the Sudan, under the auspices of the Oriental Exploration Fund of the University of Chicago, Egyptian Section. In 1919, with the financial support of John D. Rockefeller Jr., James Henry Breasted founded the Oriental Institute at the University of Chicago, as a research center for the study of the ancient Near East. For the first five years, the Oriental Institute was supported by a modest annual grant from Rockefeller; with the great gifts given later by the Rockefeller Foundations, the Oriental Institute became the leading Egyptological research center in the Western Hemisphere.

On April 25, 1923, James Henry Breasted became the first "archaeologist" to be elected to membership in the National Academy of Sciences, a personal honor that helped significantly to legitimize the struggling profession of archaeology in American academic circles. Breasted's vision established three related types of research at the Oriental Institute: archaeological fieldwork and excavation; salvage and epigraphic recording of standing monuments for publication; and the interpretation of recovered records for philological purposes and basic reference works, such as dictionaries and grammatical studies.

On June 7, 1935, Breasted married Imogen Hart Richmond (1885–1961), the divorced younger sister of his late wife Frances. James Henry Breasted died of a streptococcic infection in New York City on December 2, 1935. His remains were cremated and subsequently interred in the Breasted family plot in Rockford, Illinois, beneath a granite marker imported from Aswan, Egypt. Breasted was the real founder of professional Egyptology in the Western Hemisphere and, with George A. Reisner, one of the leading American Egyptologists of his day. During his lifetime, he acquired many distinctions, academic and otherwise.

Daniel David Luckenbill (1881–1927), American Assyriologist, was born near the borough of Hamburg in Berks County, Pennsylvania, on June 21, 1881, the son of the Rev. Benjamin Franklin Luckenbill and his wife Mary Jane (Berger). He received his early education in public schools in Pennsylvania. In 1899, Luckenbill graduated from Lehigh (later Bethlehem) Preparatory School in Bethlehem, Pennsylvania, and enrolled in the College of the University of Pennsylvania in Philadelphia, where he earned a degree in Semitic languages (AB, 1903) and was appointed Harrison Scholar in Semitics for the academic year 1903–1904. He was subsequently awarded a Harrison Fellowship in Semitics for the years 1904–1906.

While at the University of Pennsylvania, Luckenbill studied under professors Albert T. Clay, Herman V. Hilprecht, Morris Jastrow Jr., William A. Lamberton, and Dr. Hermann Ranke. During the summer semester of 1905, he continued his studies in Egyptology under Professor Adolf Erman, who had also been James Henry Breasted's teacher.

In the summer of 1906, Luckenbill entered the University of Chicago, where he was appointed Fellow in Semitics for the academic year 1906–1907. At the University of Chicago, he studied Egyptology with Professor James Henry Breasted and Assyriology with Professor Robert Francis Harper. Luckenbill received his PhD from the University of Chicago in 1907 with a dissertation entitled "A Study of the Temple Documents from the Cassite Period." In July 1907 Luckenbill's PhD thesis was reprinted by its editor Robert Francis Harper in the *American Journal of Semitic Languages and Literatures* (volume 23, number 4, pages 280–322), and was printed simultaneously as a separate private edition published by the University of Chicago Press for distribution by the University of Chicago libraries. Luckenbill spent the remainder of his academic career at the University of Chicago (Associate in Semitics, 1907–1909; Instructor, 1909–1915; Assistant Professor, 1915–1919; Associate Professor, 1919–1923; and Professor, 1923–1927).

In the autumn of 1908, Robert Francis Harper was appointed the ninth resident annual director of the American School of Archaeology at Jerusalem (as it was known at the time), and Luckenbill joined his mentor for the academic year. During the academic year 1908–1909, Luckenbill produced a visual record of his travels — approximately 500 black-and-white photographic images of the Middle East, including 76 panoramas — mostly of sites in Palestine, with a handful of pictures taken in Syria and Egypt. The negatives were purchased by the Haskell Fund for the Haskell Oriental Museum in 1910 and now form one of the earliest corpuses of original Middle Eastern views in the Oriental Institute Archives. Luckenbill's prior experience as a scholar/photographer earned him the role of official photographer for the 1919–1920 University of Chicago Expedition.

On February 24, 1914, D. D. (as he was known to friends and colleagues) married Miss Florence Parker — a Chicago heiress and University of Chicago graduate in Religious Education (SB, 1900), who was more than eight years his senior — and moved into her large home at 10340 Longwood Drive in the fashionable Beverly-Morgan Park neighborhood of Chicago. After the death of Robert Francis Harper in August 1914, Luckenbill became

Curator of the Babylonian/Assyrian section of the Haskell Oriental Museum at the University of Chicago. In 1921, Luckenbill was appointed as the first editor of the Chicago Assyrian Dictionary project, which he directed until his death.

In the spring of 1927, Luckenbill and his wife sailed to England, where D. D. intended to study cuneiform texts at the British Museum for the CAD. During the sea voyage, he contracted typhoid fever. Less than three weeks before his 46th birthday, Daniel David Luckenbill died in London, England, on June 5, 1927, at a nursing home on Tavistock Square, Camden. A funeral service was held at All Souls Church, Langham Place, London, on June 9, 1927, followed by burial in St. Marylebone (now East Finchley) Cemetery.

Ludlow Seguine Bull (1886–1954), American lawyer and Egyptologist, was born in New York City on January 10, 1886, the son of eminent ophthalmologist Dr. Charles Stedman Bull (1844–1911) and Mary Eunice Kingsbury (1856–1898). Christened with the surname of his paternal grandmother's family, Ludlow graduated from the Pomfret School in Connecticut in 1903 and earned an AB at Yale in 1907. He attended Harvard Law School and received his LLB in 1910. He was admitted to the bar in the State of New York in 1911 and practiced law with the firm of Curtis, Mallet-Prevost & Colt in New York City until 1915.

At the age of 30, on the recommendation of Albert M. Lythgoe of the Metropolitan Museum of Art, Bull embarked on a second career and enrolled at the University of Chicago to take graduate courses in Egyptology with Professor James Henry Breasted during the winter and spring academic quarters of 1916 and the entire academic year of 1916/1917.

Bull enlisted in July 1917 as a Private with the Yale Mobile Hospital Unit in the US Army Medical Service Corps, American Expeditionary Forces; by 1918, he had been promoted to First Lieutenant in the Sanitary Corps, with which he served in France until the end of the War. When he left home, Bull took with him an Egyptian grammar and an Arabic chrestomathy to study during his off hours. Ludlow Bull joined Breasted in Cairo on Christmas Day, December 25, 1919, as the first member of the Oriental Institute reconnaissance expedition to link up with its leader.

Bull wrote his dissertation, entitled "The Religious Texts from an Egyptian Coffin of the Middle Kingdom" based on study carried out in Egypt during the 1919–1920 expedition and he received his PhD from the Department of Oriental Languages and Literatures of the University of Chicago in 1922. During the winter of 1922/1923, he assisted Breasted and Dr. Alan H. Gardiner in the earliest stages of their work on the Middle Kingdom coffins in the Egyptian Museum, Cairo — the first field season of the Egyptian Coffin Texts Project. Bull's chief contribution was in listing the Cairo coffins and in classifying the materials from Pierre Lacau's personal research on the coffins.

Dr. Bull's working career as an Egyptologist is associated primarily with his alma mater, Yale University, and with the Metropolitan Museum of Art in New York. From 1925 until his death, Bull served as Honorary Curator of Egyptian Art at the Yale University Art Gallery. He was lecturer in Egyptology at Yale University, 1925–1936, and then a research associate with the rank of Professor after 1936. He was Assistant Curator in the Department of Egyptian Art of the Metropolitan Museum of Art, New York, from 1922 to 1928, and Associate Curator from 1928 until his death; he wrote a number of articles in *The Metropolitan Museum of Art Bulletin* and *The Metropolitan Museum Studies*, and was a member of the editorial board of the latter.

Ludlow Bull was Recording Secretary of the American Oriental Society, 1925–1936; Vice-president, 1938; and President, 1939.

Ludlow Seguine Bull died suddenly in his summer home on July 1, 1954, and was buried in Litchfield, Connecticut. Later that same year, Bull's estate donated his papers to the Manuscripts and Archives Department in the Sterling Memorial Library at his Alma Mater, Yale University. The earliest material in the Bull Papers at Yale dates to 1923.

William Franklin Edgerton (1893–1970), American Egyptologist and Demoticist, was born in Binghamton, Broome County, New York, on September 30, 1893, the youngest of the three sons of statistician and economist Charles Eugene Edgerton (1861–1932) and his wife Anne Benedict (White). William F. Edgerton graduated in 1911 from Central High School in Washington, DC, and studied Semitic languages at Cornell University (AB, 1915) in Ithaca, New York, where he was elected to the Cornell Chapter of Phi Beta Kappa from the junior class in 1914.

Edgerton was admitted to the University of Chicago for graduate studies and received fellowships for the years 1915 to 1918 and also studied briefly at the University of Pennsylvania. On May 22, 1918, he married fellow Cornell graduate (AB, 1912; MA, 1913) Jean Daniel Modell (1888–1980) and served in the medical department of the U.S. Army in 1918–1919. Edgerton was the youngest member of the first field expedition of the Oriental Institute of the University of Chicago, 1919–1920, under the direction of Professor James Henry Breasted. Upon returning to Chicago, Edgerton held fellowships in the Department of Oriental Languages and Literatures, 1920–1922. His PhD thesis, "Ancient Egyptian Ships and Shipping," written under Breasted, was reprinted in 1923 in the *American Journal of Semitic Languages and Literatures* (volume 39, number 2, pages 109–35), and was printed simultaneously as a separate private edition published by the University of Chicago Press for distribution by the University of Chicago libraries.

Edgerton served as an Assistant in the Oriental Institute, 1922–1923. He spent a year of post-graduate studies at Columbia University, 1923–1924. After two years of teaching experience away from the University of Chicago (Assistant Professor of Ancient History, University of Louisville, Louisville, Kentucky, 1924–1925; Associate Professor of History, Vassar College, 1925–1926), Edgerton was appointed Epigrapher with the Epigraphic Survey of the Oriental Institute of the University of Chicago, Luxor, Egypt, for three field seasons under Harold Hayden Nelson, 1926–1929. In 1927, Edgerton studied Demotic Egyptian at the University of Munich, under Professor Wilhelm Spiegelberg. Most of the remainder of his academic career was spent at the University of Chicago (Associate Professor of Egyptology, University of Chicago, 1929–1937; Professor of Egyptology, University of Chicago, 1937–1959; Chairman of the Department of Oriental Languages and Literatures, University of Chicago, 1948–1954). Edgerton received a Fulbright Grant in 1951 for a year of study at Cambridge University, England. On February 22, 1957, Linetta Margaret Cooper of Chicago became the 2nd Mrs. William Franklin Edgerton. After his retirement from the faculty of the University of Chicago in 1959, Edgerton spent two years as a Visiting Professor at the University of California, Berkeley, 1965–1967.

Edgerton collected materials for a dictionary of Demotic Egyptian — the dream of his mentor Wilhelm Spiegelberg — but it fell to Edgerton's student George R. Hughes and to Hughes' student Janet H. Johnson to make the Chicago Demotic Dictionary a reality.

After a long illness, William Franklin Edgerton died in a convalescent home in Bridgeview, Cook County, Illinois, on March 20, 1970, at the age of 76. Edgerton's personal library was bequeathed to the Oriental Institute, where it forms an important part of the older Egyptological titles in the Director's Library/Research Archives, and his professional papers and photographs are now in the Oriental Institute Archives.

Sources: Bierbriar, *Who Was Who in Egyptology*, p. 137; Hughes, "To the Members and Friends of the Oriental Institute," pp. 5–6.

William Arthur Shelton (1875–1959), son of Leroy Shelton (1835–1895) and his wife Sarah Elizabeth Rogers (1850–1896), was born in Azusa, Los Angeles County, California, on September 6, 1875. He grew up in Texas and Oklahoma. William Shelton attended Yale University, where he earned BD and MA degrees. In 1914, Shelton received an honorary Doctor of Divinity degree from Emory College (now University) in Atlanta, Georgia, and was appointed to the faculty of the new Candler School of Theology, where he served as Professor of Hebrew and Old Testament Literature until his retirement in 1930. In 1915–1916, Shelton was granted a leave of absence from Emory to continue his post-graduate studies at the University of Chicago, where he studied with Professor James Henry Breasted, among others. In 1919–1920, Shelton accompanied the inaugural reconnaissance expedition of the Oriental Institute of the University of Chicago, under the direction of Breasted, to Egypt, Mesopotamia (Iraq),

Syria, Lebanon, and Palestine. With funds made available by Georgia businessman John A. Manget, Shelton purchased antiquities for the Emory University Museum (now the Michael C. Carlos Museum) in Atlanta, Georgia. Toward the end of 1920, Shelton presented to Breasted a set of contact prints of 209 photographic images from the black-and-white negatives that he made while on the trip (now Accession 290, dated January 4, 1921, in the Oriental Institute Photographic Archives). Shelton's travel experiences as a member of the 1919–1920 Oriental Institute expedition were published as *Dust and Ashes of Empire*. William Arthur Shelton died on February 22, 1959, in Birmingham, Alabama, and is buried in Westview Cemetery, Atlanta, Georgia.

Sources: Shelton, *Dust and Ashes of Empires*; Bowen, *The Candler School of Theology: Sixty Years of Service*, pp. 172–73; Beierle, "One Brick from Babylon," pp. 8–17.

SELECTED BIBLIOGRAPHY

Abu-Manneh, Butrus. "The Christians between Ottomanism and Syrian Nationalism: The Ideas of Butrus Al-Bustani." *International Journal of Middle East Studies* 11/3 (1980): 287–304.

Ayalon, Ami. *The Press in the Arab Middle East – A History.* New York: Oxford University Press, 1995.

al-Baghdadi, Nadia. "The Cultural Function of Fiction: From Bible Translation to Libertine Literature - Social Critique Historical Criticism in Ahmad Faris ash-Shidyaq." *Arabica* 46 (1999): 375–401.

Bator, Paul. *The International Trade in Art.* Chicago: University of Chicago Press, 1983.

Bernhardsson, Magnus T. *Reclaiming a Plundered Past: Archaeology and Nation Building in Modern Iraq.* Austin: University of Texas Press, 2005.

Beierle, Andrew W. M. "One Brick from Babylon." Emory Magazine (October 1988): 8–17.

Bierbriar, Morris (editor). *Who Was Who in Egyptology.* Third revised edition. London: The Egypt Exploration Society, 1995.

Bowen, Boone M. *The Candler School of Theology: Sixty Years of Service.* Atlanta: Emory University, 1974.

Breasted, Charles. *Pioneer to the Past: The Story of James Henry Breasted, Archaeologist, Told by His Son, Charles Breasted.* New York: Charles Scribner's Sons, 1943.

Breasted, James H. *Al-'Uṣūr al-qadīma.* Translation of *Ancient Times* into Arabic by Da'ud Qurban. Beirut: The American Press, 1926. Second edition, 1930.

——. *Ancient Records of Egypt: Historical Documents from the Earliest Times to the Persian Conquest.* Five volumes. Chicago: University of Chicago Press, 1906–07.

——. *Ancient Times: A History of the Early World; An Introduction to the Study of Ancient History and the Career of Early Man.* Boston: Ginn, 1916.

——. *The Battle of Kadesh: A Study in the Earliest Known Military Strategy.* Decennial Publications 5. Chicago: University of Chicago Press, 1903.

——. *A History of Egypt from the Earliest Times to the Persian Conquest.* New York: Charles Scribner's Sons, 1905.

——. *The Oriental Institute.* University of Chicago Survey 12. Chicago: University of Chicago Press, 1933.

——. Report of the First Expedition of the Oriental Institute of the University of Chicago. Undated (ca. 1920) manuscript in the Oriental Institute Archives; reprinted in its entirety in *Appendix B.*

Calverley, Amice M., and Alan H. Gardiner. *The Temple of King Sethos I at Abydos,* Volume I: *The Chapels of Osiris, Isis and Horus.* London: Egypt Exploration Society; Chicago: University of Chicago Press, 1933.

Chatterjee, Partha. *The Nation and Its Fragments: Colonial and Postcolonial Histories.* Princeton: Princeton University Press, 1993.

Corbett, Elena Dodge. Jordan First: A History of the Intellectual and Political Economy of Jordanian Antiquity. Unpublished Ph.D. dissertation, University of Chicago, Department of Near Eastern Languages and Civilizations, June 2009.

Cuno, James. *Who Owns Antiquity? Museums and the Battle Over Our Ancient Heritage.* Princeton: Princeton University Press, 2008.

Dakhli, Leyla. *Une génération d'intellectuels arabes: Syrie et Liban, 1908–1940.* Paris: Karthala, 2009.

Davies, Nina M., and Alan H. Gardiner. *Ancient Egyptian Paintings,* Volume I. Chicago: University of Chicago Press, 1936.

Dodge, Toby. *Inventing Iraq: The Failure of Nation-Building and a History Denied.* New York: Columbia University Press, 2003.

Fagan, Brian. *Rape of the Nile: Tomb Robbers, Tourists, and Archaeologists in Egypt.* Revised edition. New York: Basic Books, 2004.

Elshakry, Marwa. "The Gospel of Science and American Evangelism in Late Ottoman Beirut." *Past and Present* 196/1 (2007): 173–214.

Faraj, Nadia. "The Lewis Affair and the Fortunes of al-Muqtataf." *Middle East Studies* 7 (1972): 73–83.

Fromkin, David. *A Peace to End All Peace: The Fall of the Ottoman Empire and the Creation of the Modern Middle East.* New York: Holt, 2009.

Garstang, John. "Eighteen Months Work of the Department of Antiquities for Palestine, July 1920–December 1921." *Palestine Exploration Fund Quarterly Statement* (April 1922): 58.

Gershoni, Israel, and James P. Jankowski. *Egypt, Islam, and the Arabs: The Search for Egyptian Nationhood, 1900–1930.* Studies in Middle Eastern History. New York: Oxford University Press, 1987.

Gerstenblith, Patty. "Schultz and Barakat: Universal Recognition of National Ownership of Antiquities." *Art, Law and Antiquity* 14 (2009): 21–48.

Gelvin, James L. "The Ironic Legacy of the King-Crane Commission." In *The Middle East and the United States: A Historical and Political Reassessment*, edited by David W. Lesch, pp. 13–29. Fourth edition. Boulder: Westview Press, 2007.

——. *The Israel-Palestine Conflict: One Hundred Years of War*. New edition. Cambridge: Cambridge University Press, 2007.

——. *The Modern Middle East: A History*. Second edition. New York: Oxford University Press, 2008.

Gibson, Shimon. "British Archaeological Institutions in Mandatory Palestine, 1917–1948." *Palestine Exploration Quarterly* 131 (1999): 115–43.

Glaß, Dagmar. *Der Muqtataf und seine Öffentlichkeit: Aufklärung, Räsonnement und Meinungsstreit in der frühen arabischen Zeitschriftenkommunikation*. Würzburg: Ergon, 2004.

Goldschmidt, Arthur, Jr. *Modern Egypt: The Formation of a Nation State*. Second edition. Boulder: Westview Press, 2004.

Goode, James F. *Negotiating for the Past: Archaeology, Nationalism and Diplomacy in the Middle East 1919–1941*. Austin: University of Texas Press, 2007.

Hourani, Albert Habib. *Arabic Thought in the Liberal Age, 1798–1939*. London and New York: Oxford University Press, 1962.

Hughes, George R.. "To the Members and Friends of the Oriental Institute." *Oriental Institute Report 1969/70*, pp. 1–11.

Jasanoff, Maya. *Edge of Empire: Lives, Culture and Conquest in the East 1750–1850*. New York: Random House/Vintage Press, 2005.

Kaufman, Asher. *Reviving Phoenicia: In Search of Identity in Lebanon*. London: I. B. Tauris, 2004.

Khoury, Philip S. *Syria and the French Mandate: The Politics of Arab Nationalism, 1920–1945*. Princeton: Princeton University Press, 1987.

Kletter, Raz. *Just Past? The Making of Israeli Archaeology*. London: Equinox Publishing, 2006.

Kuklick, Bruce. *Puritans in Babylon: The Ancient Near East and American Intellectual Life, 1880–1930*. Princeton: Princeton University Press, 1996.

Liebmann, Matthew, and Uzma Z. Rizvi (editors). *Archaeology and the Postcolonial Critique*. Lanham: AltaMira Press, 2008.

Lloyd, Seton. *Foundations in the Dust: The Story of Mesopotamian Exploration*. Revised edition. New York: Thames & Hudson, 1980.

Makdisi, Saree. "Postcolonial Literature in a Neocolonial World: Modern Arabic Culture and the End of Modernity." *Boundary* 22/1 (1995): 85–115.

Makdisi, Ussama. *Artillery of Heaven: American Missionaries and the Failed Conversion of the Middle East*. Ithaca: Cornell University Press, 2008.

Marchand, Suzanne L. *Down from Olympus: Archaeology and Philhellenism in Germany, 1750–1970*. Princeton: Princeton University Press, 1996.

Merryman, John Henry (editor). *Imperialism, Art and Restitution*. Cambridge: Cambridge University Press, 2006.

Oren, Michael B. *Power, Faith, and Fantasy: America in the Middle East, 1776 to the Present*. New York: W. W. Norton, 2007.

Philipp, Thomas D. *The Autobiography of Jurji Zaidan: Including Four Letters to His Son*. Washington, DC: Three Continents Press, 1990.

Reid, Donald M. *The Odyssey of Farah Anṭūn: A Syrian Christian Quest for Secularism*. Studies in Middle Eastern History 2. Minneapolis: Bibliotheca Islamica, 1975.

——. "The Syrian Christians and Early Socialism in the Arab World." *International Journal of Middle East Studies* 5/2 (1974): 177–93.

——. "Syrian Christians, the Rags-to-Riches Story, and Free Enterprise." *International Journal of Middle East Studies* 1/4 (1970): 358–67.

——. *Whose Pharaohs? Archaeology, Museums, and Egyptian National Identity from Napoleon to World War I*. Berkeley: University of California Press, 2002.

Russell, John. *From Nineveh to New York: The Strange Story of the Assyrian Reliefs in the Metropolitan Museum and the Hidden Masterpiece at Canford School*. New Haven: Yale University Press, 1997.

Said, Edward W. *Orientalism*. New York: Pantheon Books, 1978.

Satia, Priya. *Spies in Arabia: The Great War and the Cultural Foundations of Britain's Covert Empire in the Middle East*. Oxford: Oxford University Press, 2008.

Shaw, Wendy K. *Possessors and Possessed: Museums, Archaeology and the Visualization of History in the Late Ottoman Empire*. Berkeley: University of California Press, 2003.

Shelton, William Arthur. *Dust and Ashes of Empires*. Nashville: Publishing House of the Methodist Episcopal Church, South, 1922; Cokesbury Press, 1924.

Silberman, Neil A. *Between Past and Present: Archaeology, Ideology, and Nationalism in the Modern Middle East.* New York: H. Holt, 1989.

——. *Digging for God and Country: Exploration, Archeology, and the Secret Struggle for the Holy Land, 1799–1917.* New York: Alfred A. Knopf, 1982.

——. "Power, Politics and the Past: The Social Construction of Antiquity in the Holy Land." In *The Archaeology of Society in the Holy Land*, edited by Thomas E. Levy, pp. 9–23. London: Leicester University Press, 1995.

Teeter, Emily. "Egypt in Chicago: A Story of Three Collections." In *Egyptian Studies*, edited by Zahi Hawass and Jennifer Wegner. Cairo: Supreme Council of Antiquities, 2010.

Thomas, Thelma K. *Dangerous Archaeology: Francis Willey Kelsey and Armenia (1919–1920).* Ann Arbor: Kelsey Museum, 1990.

Tripp, Charles. *A History of Iraq.* Third edition. Cambridge: Cambridge University Press, 2007.

Trümpler, Charlotte (editor). *Das grosse Spiel: Archäologie und Politik zur Zeit des Kolonialismus (1860–1940).* Cologne: DuMont, 2008.

Urice, Stephen K. "The Beautiful One Has Come — To Stay." In *Imperialism, Art and Restitution*, edited by J. H. Merryman, pp. 135–75. New York: Cambridge University Press, 2006.

Wilson, Karen L. *Bismaya: Recovering the Lost City of Adab.* Chicago: The Oriental Institute, in press.

Wilson, Mary C. *King Abdullah, Britain, and the Making of Jordan.* Cambridge: Cambridge University Press, 1990.

Ziadat, Adel A. *Western Science in the Arab World: The Impact of Darwinism, 1860–1930.* London: MacMillan, 1986.

CONCORDANCE OF OBJECTS IN THE EXHIBIT

Registration Number	Figure Number	Description
EGYPTIAN OBJECTS FROM THE ORIENTAL INSTITUTE		
OIM 9864B	4.3	Inlay
OIM 9864D	4.3	Inlay
OIM 9866B	4.3	Inlay
OIM 9868C	4.3	Inlay
OIM 10101	4.4	Model Water Clock
OIM 10480	4.18	Cylinder Seal
OIM 10486B	4.6	Book of the Dead (Papyrus Milbank)
OIM 10517A	4.15	Fragment of Composite Statue
OIM 10517B	4.15	Fragment of Composite Statue
OIM 10548	4.13	Battle Axe
OIM 10584	4.12	Figure of Amun
OIM 10626	4.9	Serving Statue
OIM 11211	4.10	Flint
OIM 11219	4.10	Flint
MESOPOTAMIAN OBJECTS FROM THE ORIENTAL INSTITUTE		
OIM A2638	4.11	Tablet
OIM A2645	4.11	Tablet
OIM A2651	4.11	Tablet
OIM A2655	4.11	Tablet
OBJECTS FROM THE ART INSTITUTE OF CHICAGO		
1920.252	4.16	Bronze Jackal
1920.262	4.17	Wall Fragment of Amenemhet and Hemet

INDICES

GENERAL INDEX

INDEX OF PERSONAL NAMES

INDEX OF GEOGRAPHICAL NAMES